Effects of Activated Carbon Characteristics on Organic Contaminant Removal

The mission of the Awwa Research Foundation (AwwaRF) is to advance the science of water to improve the quality of life. Funded primarily through annual subscription payments from over 1,000 utilities, consulting firms, and manufacturers in North America and abroad, AwwaRF sponsors research on all aspects of drinking water, including supply and resources, treatment, monitoring and analysis, distribution, management, and health effects.

From its headquarters in Denver, Colorado, the AwwaRF staff directs and supports the efforts of over 500 volunteers, who are the heart of the research program. These volunteers, serving on various boards and committees, use their expertise to select and monitor research studies to benefit the entire drinking water community.

Research findings are disseminated through a number of technology transfer activities, including research reports, conferences, videotape summaries, and periodicals.

Effects of Activated Carbon Characteristics on Organic Contaminant Removal

Prepared by:
Detlef R.U. Knappe, Lei Li,
Patricia A. Quinlivan, and **Travis B. Wagner**
Department of Civil Engineering
North Carolina State University
Raleigh, NC 27695-7908

Sponsored by:
Awwa Research Foundation
6666 West Quincy Avenue
Denver, CO 80235-3098

Published by the

Awwa Research Foundation
Advancing the Science of Water®

IWA Publishing

DISCLAIMER

This study was funded by the Awwa Research Foundation (AwwaRF). AwwaRF assumes no responsibility for the content of the research study reported in this publication or for the opinions or statements of fact expressed in the report. The mention of trade names for commercial products does not represent or imply the approval or endorsement of AwwaRF. This report is presented solely for informational purposes.

Published by IWA Publishing, Alliance House, 12 Caxton Street, London SW1H 0QS, UK

Telephone: +44 (0) 20 7654 5500; Fax: +44 (0) 20 7654 5555; Email: publications@iwap.co.uk;
Web: **www.iwapublishing.com**

AwwaRF report number 90926F. Originally published by AwwaRF for its subscribers in 2003.
IWA Publishing version published 2004.

ISBN 1 84339 841 9

CONTENTS

TABLES

FIGURES

FOREWORD

The Awwa Research Foundation is a nonprofit corporation that is dedicated to implementation of a research effort to help utilities respond to regulatory requirements and traditional high-priority concerns of the industry. The research agenda is developed through a process of consultation with subscribers and drinking water professionals. Under the umbrella of a Strategic Research Plan, the Research Advisory Council prioritizes the suggested projects based on current and future research needs, applicability, and past work; the recommendations are forwarded to the Board of Trustees for final selection. The foundation also sponsors research projects through the unsolicited proposal process; the Collaborative Research, Research Applications, and Tailored Collaborations programs; and various joint research efforts with organizations such as the U.S. Environmental Protection Agency, the U.S. Bureau of Reclamation, and the Association of California Water Agencies.

This publication is a result of one of these sponsored studies, and it is hoped that its findings will be applied in communities throughout the world. The following report serves not only as a means of communicating the results of the water industry's centralized research program but also as a tool to enlist the further support of the nonmember utilities and individuals.

Projects are managed closely from their inception to the final report by the foundation's staff and large cadre of volunteers who willingly contribute their time and expertise. The foundation serves a planning and management function and awards contracts to other institutions such as water utilities, universities, and engineering firms. The funding for this research effort comes primarily from the Subscription Program, through which water utilities subscribe to the research program and make an annual payment proportionate to the volume of water they deliver, and consultants and manufacturers subscribe based on their annual billings. The program offers a cost-effective and fair method of funding research in the public interest.

A broad spectrum of water supply issues is addressed by the foundation's research agenda: resources, treatment and operations, distribution and storage, water quality and analysis, toxicology, economics, and management. The ultimate purpose of the coordinated effort is to assist water suppliers to provide the highest possible quality of water economically and reliably. The true benefits are realized when the results are implemented at the utility level. The foundation's trustees are pleased to offer this publication as a contribution toward that end.

Edmund G. Archuleta, P.E.
Chair, Board of Trustees
Awwa Research Foundation

James F. Manwaring, P.E.
Executive Director
Awwa Research Foundation

ACKNOWLEDGMENTS

The authors of this report would like to thank the following people and organizations for their cooperation, participation, and support of this project:

- AwwaRF for funding this project
- Misha Hasan and Chris Rayburn at AwwaRF for their excellent support as project managers
- The project advisory committee (PAC) consisting of Laura Cummings, Fran DiGiano, and Gayle Newcombe for their helpful comments throughout the course of this research
- The National Science Foundation (NSF) for supporting Patricia A. Quinlivan through a Graduate Research Fellowship
- Larry McCollum and other dedicated employees at the Contra Costa Water District for providing the research team at NC State University with Sacramento–San Joaquin Delta water
- Nippon Kynol for donating activated carbon fibers and Professor Yoshihiko Matsui at Gifu University, Japan, for obtaining the activated carbon fibers from Nippon Kynol and shipping them to the research team at NC State University
- Calgon Carbon Corporation and PicaUSA for donating granular activated carbons
- David Black in the Environmental Engineering Research Laboratory at NC State University for his high-quality gas chromatography work
- Beverly Taylor in the Soil Science Department at NC State University for CHN analyses

EXECUTIVE SUMMARY

Introduction

Many water treatment plants need to remove objectionable trace organic compounds, and activated carbon adsorption is often the best available technology. However, utilities face the challenge of having to choose from a large variety of activated carbons, and in many cases, iodine number or BET surface area values are utilized in the selection process. Although neither parameter correlates well with adsorption capacities for trace organic contaminants, alternative activated carbon selection criteria that are based on fundamental adsorbent and adsorbate properties are lacking to date.

Objectives and Approach

The objective of this research was to explore alternative activated carbon selection criteria by determining the effects of activated carbon pore structure and surface chemistry on the adsorption of a relatively polar (methyl tertiary-butyl ether = MTBE) and a relatively nonpolar (trichloroethene = TCE) trace organic compound in the presence of natural organic matter (NOM). To systematically study the effects of activated carbon pore structure and surface chemistry on adsorption, a matrix of activated carbon fibers (ACFs) was examined. The ACF matrix contained three activation levels (ACF10, ACF15, and ACF20) and four surface chemistry levels (acid-washed, oxidized, hydrogen-treated, ammonia-treated). ACFs served as model adsorbents that permitted a relatively independent evaluation of the effects of surface chemistry and pore structure on the adsorption of trace organic compounds from drinking water. However, the cost of ACFs is currently too high to compete with traditional granular and powdered activated carbons for water treatment. To verify that trends established with ACFs are also valid for traditional activated carbons, three commercially available granular activated carbons (GACs) were studied as well [coal-based Filtrasorb 600 (F600), coconut shell-based G219, and wood-based Picazine]. Physical and chemical adsorbent characterization included (1) pore volume, pore size distribution, and surface area determinations by ultra-low pressure N_2 adsorption, (2) point of zero charge (PZC) determinations by mass titrations, (3) elemental analyses, and (4) determination of surface functional groups by Fourier Transform Infrared (FTIR) spectroscopy. Batch adsorption isotherm data for MTBE and TCE were collected in ultrapure water and in San Joaquin Delta Water (SJDW, Concord, CA).

Results

Adsorbent Characterization

BET surface areas for acid-washed ACFs, which served as the starting materials for subsequent chemical modifications, increased with increasing degree of activation from 760 m^2/g for ACF10 to 1670 m^2/g for ACF20. Chemical modification techniques changed BET surface areas by less than 10% compared to the corresponding acid-washed ACF. Oxidative treatments reduced micropore volumes of the ACFs by about 12 to 23% while heat treatments in H_2 or NH_3 atmospheres had little effect on micropore volumes (4% decrease to 7% increase). With the exception of oxidized ACF15, mesopore volume changes were in the ±20% range and were relatively small (in terms of cm^3/g) given the microporous nature of the ACFs. Micropore size distributions were calculated with the density functional theory (DFT), which showed that pores of acid-washed ACF10 were concentrated in the 7-11 Å diameter range while those of acid-washed ACFs 15 and 20 were concentrated in the 9-13 Å diameter range. As was observed with the BET surface area data, chemical treatments had a negligible effect on pore size distributions. Thus, the developed ACF matrix permitted the evaluation of surface chemistry effects on trace organic compound adsorption in the absence of pore structure effects.

With respect to surface chemistry, acid-washed ACF surfaces exhibited a slightly acidic character with the point of zero charge (PZC) occurring at about pH 5. ACF oxidation by H_2O_2 lowered the pH_{PZC}, consistent with the appearance of carboxylic functional groups in FTIR spectra. Hydrogen-treatment created ACFs that were more basic than acid-washed ACFs, and the ammonia-treated ACFs assumed a basic character. FTIR and elemental analyses indicated that the increased basicity of hydrogen- and ammonia-treated ACFs was in part due to the removal of oxygen-containing functional groups. The results of elemental analyses and mass titrations indicated that a given chemical treatment yielded ACFs with similar heteroatom content and surface acidity/basicity. Thus, the ACF matrix also permitted the study of pore structure effects on trace organic compound adsorption with little interference from surface chemistry effects.

Commercial GACs exhibited pore size distributions that were broader and less uniform than those of the ACFs. Micropores of F600 were concentrated in the 7 to 12 Å width range while those of G219 were concentrated in the 7 to 13 Å width range. The micropores of Picazine were largest among the commercially available GACs and were concentrated in the 9 to 15 Å width range. F600 and G219, GACs with oxygen contents of about 3 to 5% by weight, exhibited a basic character, while Picazine (~ 16% oxygen by weight) was strongly acidic.

Single-Solute Isotherms

Single-solute MTBE and TCE isotherm data were collected for the entire ACF matrix. To quantitatively compare single-solute isotherm results, data were described by the Freundlich model [$q_e = K (C_e)^{1/n}$], where q_e and C_e are the equilibrium solid-phase and liquid-phase concentrations, respectively, and K and 1/n are isotherm constants that represent adsorption capacity and strength, respectively. As expected, adsorption capacities for hydrophobic TCE were greater than for the more hydrophilic MTBE. Regarding pore structure, correlations showed that TCE adsorbed primarily in pores with widths in the 7-10 Å range while MTBE adsorbed primarily in pores with widths in the 8-11 Å range. These pore widths corresponded to approximately 1.3 to 1.8 times the kinetic diameter of the target molecules (5.6 Å for TCE, 6.2 Å for MTBE). The results also illustrated that the BET surface area was not a suitable adsorbent selection criterion; e.g., the TCE adsorption capacity of ACF 10 was largest even though its BET surface area was about 50% smaller than that of ACFs 15 and 20. However, ACF 10 contained a larger pore volume in the 7-10 Å width range.

At a given activation level, hydrogen-treated ACFs, the most hydrophobic adsorbents in the ACF matrix, exhibited the largest MTBE and TCE adsorption capacities while oxidized ACFs, the most polar adsorbents in the ACF matrix, exhibited the lowest. To evaluate surface chemistry effects in the absence of pore structure effects, adsorption capacities at an equilibrium liquid-phase concentration of 50 μg/L ($q_{e, 50}$) were normalized by pore volume in the appropriate size range (7-10 Å for TCE, 8-11 Å for MTBE). The thus normalized MTBE and TCE adsorption capacities increased with increasing adsorbent hydrophobicity (i.e., with decreasing O+N content of the activated carbon). For adsorbents with a given pore size distribution, the largest adsorption capacities will therefore be realized with hydrophobic activated carbons, regardless of whether the targeted compound is hydrophobic (TCE) or hydrophilic (MTBE) in character. This result can be explained by the preferential adsorption of water on polar surface sites.

Isotherms in San Joaquin Delta Water (SJDW)

When conducting isotherm experiments in the presence of NOM, pore size requirements for TCE and MTBE adsorption were similar to those determined from single-solute isotherms (7-9 Å for TCE, 8-11 Å for MTBE). However, NOM adsorption severely reduced MTBE adsorption capacities by pore blockage when the upper end of the required pore size range (11 Å for MTBE) closely matched the upper end of the dominant pore width range of the adsorbent (11 Å for ACF10). In contrast, pore blockage was not important for ACFs, in which the micropore size distribution extended to widths that corresponded to about twice the kinetic diameter of the target adsorbate. Regarding surface chemistry, hydrophobic carbons outperformed hydrophilic carbons in a fashion similar to that observed in single-solute systems. These results further substantiate the need to choose hydrophobic adsorbents with a proper pore size distribution for the removal of both hydrophobic and hydrophilic trace organic pollutants from drinking water sources.

To illustrate the importance of adsorbent selection, adsorbents were ranked in terms of the carbon dosage required to remove 90% of either TCE or MTBE from SJDW. To achieve 90% TCE removal from SJDW, adsorbent doses of about 8 to 300 mg/L were required, and the rankings were affected both by adsorbent pore structure and surface chemistry. The most suitable adsorbents for TCE were hydrophobic adsorbents that exhibited a large pore volume in the 7 to 9 Å width range; i.e., acid-washed, hydrogen-treated, and ammonia-treated ACF10. Among the commercially available GACs, F600 most closely met the requirements for a suitable adsorbent for TCE removal from natural water, and this adsorbent was nearly 20 times as effective as the polar wood-based Picazine. At an initial TCE concentration of 100 μg/L in SJDW, competing NOM decreased TCE adsorption capacities by about 30 to 50% compared to the corresponding single-solute TCE adsorption capacities, and the magnitude of the capacity reduction showed little dependence on pore structure or surface chemistry. Because the effects of pore structure and surface chemistry had little effect on the competitive effect of NOM, the adsorbent rankings for TCE removal from SJDW were therefore similar to those for TCE removal in single-solute systems.

To achieve 90% MTBE removal from SJDW, adsorbent doses of about 160 to 8000 mg/L were required. MTBE adsorption was most effective with hydrophobic adsorbents that exhibited a large pore volume in the 8 to 11 Å width range and whose micropore widths extended to about 13 Å to prevent pore blockage. Thus, the most effective adsorbents for MTBE were hydrophobic ACFs 15 and 20 (i.e., hydrogen-treated, acid-washed, and ammonia-treated) whose pores were concentrated in the 9-13 Å width range. Among the commercially available GACs, coconut-shell-based G219 was most effective for MTBE removal from natural water (dominant pore widths: 7-13 Å) while the polar Picazine GAC was again the least effective. In terms of capacity reduction percentage, NOM adsorption had the most detrimental effects on the MTBE adsorption capacities of acid-washed and ammonia-treated ACF10, a result that was primarily attributable to pore blockage or pore constriction as described above. While capacity reduction percentages were lowest for polar adsorbents with larger micropores (i.e., Picazine and oxidized ACF 15 and 20), a result consistent with the electrostatic repulsion between negatively charged adsorbent surfaces and negatively charged NOM, polar adsorbents still performed poorly in natural water because of the detrimental effects of enhanced water adsorption. Rankings of the adsorbents in terms of MTBE removal effectiveness from SJDW were generally similar to those observed in ultrapure water with two notable exceptions; the effectiveness of both acid-washed and ammonia-treated ACF10 decreased greatly because of the detrimental effects of pore blockage/constriction on MTBE adsorption capacity.

Conclusions

Both adsorbent surface chemistry and pore structure affect the adsorption capacity for trace organic compounds from drinking water sources. To maximize adsorption capacity, the following factors should be considered when selecting activated carbons:

1. Regarding pore size distribution, the dominant micropore width should correspond to about 1.5 times the kinetic diameter of the target molecule, and the micropore size distribution should extend to widths that correspond to approximately twice the kinetic diameter of the target molecule to prevent pore blockage by NOM.
2. Regarding surface chemistry, hydrophobic adsorbents with an O+N content of <2-3 mmol/g should be selected.

The results of this study clearly showed that the BET surface area is an inappropriate criterion for selecting effective adsorbents for the removal of trace organic contaminants from drinking water sources. E.g., acid-washed ACF10 with a BET surface area of 760 m^2/g exhibited a TCE adsorption capacity that was about 3 times larger than that of acid-washed ACF20 with a BET surface area of 1670 m^2/g. Also, among the commercial GACs that were evaluated in this study, the GAC with the largest BET surface area exhibited the lowest adsorption capacity for TCE and MTBE because its micropores were too large and its surface too polar. Similarly, the iodine number, which correlates well with BET surface area, is an inappropriate activated carbon selection criterion.

While a large adsorption capacity is important for selecting effective activated carbons, rapid adsorption kinetics can be of even greater importance when contact times are short (e.g., Newcombe et al., 2002). In that case, the mesopore volume of an activated carbon should become an additional selection criterion. Recommendations regarding a minimum mesopore volume are not yet possible because adsorption kinetics were not measured in this study.

Overall, the results of this study emphasize that selection criteria such as the unit cost of activated carbon (e.g., \$/ton), iodine number, or BET surface area (e.g., \$/(mg/g) of iodine number or \$/(m^2/g) of BET surface area) are inappropriate for selecting an activated carbon that can meet a given treatment goal at minimum cost. Currently, it is still difficult to write bid specifications that are solely based on activated carbon pore structure and surface chemistry although the above recommendations can serve as initial guidelines. Meanwhile, utilities should require performance-based tests in bid specifications. For applications involving powdered activated carbon (PAC), for which adsorption kinetics are especially important, performance-based tests will yield the required PAC dose to meet a given treatment goal under conditions that are applicable to a given plant (contact time, mixing, sequence of chemical additions). Treatment costs can then be estimated from the results of such tests by multiplying the required PAC dose by the unit cost of the PAC [i.e., Treatment cost ($/day) = Required PAC dose to meet treatment goal (mg/L) * PAC cost ($/lb) * daily flow (mgd) * 8.34 (lb/Mgal)/(mg/L)].

CHAPTER 1
INTRODUCTION AND OBJECTIVES

Activated carbon is widely used for the control of synthetic and naturally occurring organic chemicals in drinking water. When faced with the task of selecting a suitable activated carbon for the removal of trace organic compounds from drinking water sources, water treatment professionals are faced with the problem of having to choose from a wide array of activated carbons that are marketed by numerous manufacturers. Frequently, the selection process is based on the cost of activated carbon (e.g., \$/ton) or on adsorbent descriptors such as the iodine number or the Brunauer, Emmett, and Teller (BET) surface area obtained from nitrogen adsorption at 77 K (e.g., \$/(mg/g) of iodine number or \$/(m^2/g) of BET surface area). However, the carbon usage rate and thus the cost of activated carbon adsorption processes is generally unrelated to these parameters because the molecular dimensions and chemical characteristics of iodine and nitrogen are not representative of typical organic contaminants in aqueous solution. Consequently, these parameters do not recognize the importance of molecular sieving; e.g. the exclusion of larger organic contaminants from adsorbent pores that are accessible to nitrogen or iodine (Kasaoka et al. 1989a, Kasaoksa et al. 1989b). Furthermore, effects of pore surface chemistry on adsorbate/adsorbent and solvent/adsorbent interactions (Kaneko et al. 1989, Pendleton et al. 1997, Franz et al. 2000, Considine et al. 2001) are not accounted for.

To develop a rational framework for the selection of activated carbons an improved understanding of the effects of pore size and pore surface chemistry on organic contaminant adsorption from drinking water is required. Therefore, the first objective of this study was to systematically evaluate the effects of activated carbon pore structure and surface chemistry on the adsorption of two common drinking water contaminants: the relatively polar fuel oxygenate methyl tertiary-butyl ether (MTBE) and the relatively nonpolar solvent trichloroethene (TCE). The second objective was to develop simple descriptors of activated carbon characteristics that facilitate the selection of suitable adsorbents for the removal of organic contaminants from aqueous solution.

To systematically study the effects of activated carbon pore structure and surface chemistry on adsorption, a matrix of activated carbon fibers (ACFs) was examined. The ACF matrix contained three activation levels (ACF10, ACF15, ACF20) and four surface chemistry levels (acid-washed, oxidized, hydrogen-treated, and ammonia-treated). ACFs served as model adsorbents that permitted a relatively independent evaluation of the effects of surface chemistry and pore structure on trace organic compound adsorption from drinking water. However, the cost of ACFs is currently too high to compete with traditional granular and powdered activated carbons for water treatment. To verify that trends established with ACFs are also valid for traditional activated carbons, three commercially available granular activated carbons (GACs) were studied as well: bituminous coal-based Filtrasorb 600 (F600), coconut shell-based G-219, and wood-based Picazine. Physical and chemical adsorbent characterization included (1) pore volume, pore size distribution, and surface area determinations by ultra-low pressure N_2 adsorption, (2) point of zero charge (PZC) determinations by mass titrations, (3) elemental analyses, and (4) determination of surface functional groups by Fourier Transform Infrared (FTIR) spectroscopy. Adsorption capacities were determined for natural organic matter (NOM), as measured by dissolved organic carbon (DOC) and UV absorbance, as well as for MTBE and TCE. MTBE and TCE adsorption capacities were determined from single-solute isotherm experiments and from isotherm experiments conducted in the presence of competing NOM.

CHAPTER 2
LITERATURE REVIEW

Powdered and granular activated carbons are typically manufactured from relatively heterogeneous base materials such as bituminous coal, lignite, coconut-shells, or wood while more recently developed activated carbon fibers (ACFs) are prepared from homogeneous polymeric base materials such as polyacrylonitrile, cellulose or phenolic resin (Kasaoka et al. 1989a). Following carbonization of the base material, further development of the internal pore structure is typically achieved by thermal oxidation using steam and CO_2 or chemical oxidation using phosphoric acid. Both the base material and activation conditions affect activated carbon pore structure and surface chemistry (Smisek and Cerny 1970).

ACTIVATED CARBON PORE STRUCTURE

Activated carbon is a highly porous material with an internal surface area of about 500-2500 m^2/g (Derbyshire et al. 2001). The pores that give rise to this large surface area can be envisioned as spaces between irregularly arranged graphite-like platelets (Leon y Leon et al. 1992) or condensed, polyaromatic sheets (Bansal et al. 1988), the fundamental building blocks of activated carbons. To classify pores according to size, the International Union of Pure and Applied Chemistry differentiates between (1) micropores (<2 nm width), (2) mesopores (2 - 50 nm width), and (3) macropores (> 50 nm width) (Sing et al. 1985). Powdered and granular activated carbons for water treatment typically exhibit a heterogeneous pore structure, in which micropores, mesopores and macropores are present (Kasaoka et al. 1989a). In contrast, ACFs exhibit monodispersed pore size distributions (Kasaoka et al 1989a), an observation that was confirmed by scanning tunneling microscopy (Daley et al. 1996). The latter study showed that ACF preparation could be controlled such that only ellipsoidal micropores were present in the bulk fiber. At the ACF surface, both mesopores and micropores were observed; however, the mesopores were oriented parallel to the fiber axis and thus did not penetrate the bulk fiber to a depth greater than about 60 nm. Because of their narrower pore size distribution, ACFs are particularly suitable to elucidate pore size effects on organic compound adsorption from aqueous solution (Kasaoka et al. 1989b, Pelekani and Snoeyink 1999, 2000, 2001).

The size of adsorbent pores affects the adsorption of organic contaminants in two ways. First, adsorption strength increases with decreasing pore size because (1) contact points between the adsorbate and the adsorbent surface increase (Newcombe et al. 1997) and (2) adsorption potentials between opposing pore walls begin to overlap once the micropore width is less than twice the adsorbate diameter (e.g., Dubinin 1960, Sing 1995). Second, size exclusion limits the adsorption of contaminants of a given size and shape if pores are too small. In aqueous systems, size exclusion is observed when the pore width is smaller than about 1.7 times the second largest dimension of the adsorbate (Kasaoka et al. 1989b).

The above observations suggest that adsorption takes place in the smallest pores that are accessible to a given chemical. Because many organic contaminants are small (i.e., total surface areas of about 100-300 $Å^2$ (Okouchi et al. 1992), which corresponds to spherical diameters of about 5.5-10 Å), the above observations suggest that the presence of small micropores is important for their removal from aqueous solution. For example, Newcombe et al. (2002) observed that the adsorption capacity for the naturally occurring taste and odor compound 2-methylisoborneol (MIB) correlated well with the volume of micropores with widths of about 10 to 12 Å.

ACTIVATED CARBON SURFACE CHEMISTRY

At the edges of the condensed, polyaromatic sheets that constitute the building blocks of activated carbons, heteroatoms; i.e., atoms other than carbon, are encountered that define the chemical characteristics of activated carbon surfaces. Oxygen is an important heteroatom that commonly occurs in the form of carboxylic acid groups (–COOH), phenolic hydroxyl groups (–OH), and quinone carbonyl groups (>C=O) (Boehm et al. 1964, Puri 1970, Mattson and Mark 1971, Snoeyink and Weber 1972, Leon y Leon and Radovic 1992, Boehm 1994). Activated carbons assume an acidic character when exposed to oxygen between 200 and 700°C or to oxidants such as hydrogen peroxide, nitric acid, or nitric and sulfuric acid mixtures in aqueous solution (e.g., Puri 1970, Puri 1983). The increased activated carbon acidity is primarily explained by the formation of carboxylic acid and phenolic hydroxyl groups (Leon y Leon and Radovic 1992), and X-ray photoelectron spectroscopy (XPS) data showed that the above oxidation techniques non-selectively increased the surface concentrations of both groups (Polovina et al. 1997, Mangun et al. 1999). While the majority of the oxidation conditions employed by Mangun et al. (1999) also increased the surface concentration of quinone carbonyl groups, the opposite trend was observed by Polovina et al. (1997).

The presence of basic oxygen-containing functionalities such as chromene- and pyrone-type structures has also been postulated (Garten and Weiss 1957, Boehm and Voll 1970). However, a consensus is developing that the basic character of activated carbons arises primarily from delocalized π-electrons on the condensed polyaromatic sheets (Leon y Leon et al. 1992, Barton et al. 1997). These electron-rich Lewis base sites develop as oxygen is removed from the activated carbon surface, e.g., by heat treatment in an inert atmosphere. This treatment does not only remove inherently acidic groups, but also other oxygen-containing functionalities that decrease the basicity of activated carbons by attracting and thus localizing π-electrons of the condensed, polyaromatic sheets (Coughlin and Ezra 1968). The removal of oxygen not only renders an activated carbon surface more basic but also less polar, a desirable trait when the adsorption of organic contaminants from aqueous solution is the primary treatment objective.

Heat treatment in an inert atmosphere effectively removes oxygen-containing functional groups from the activated carbon surface, but it produces highly reactive sites such as free-radical edge sites and dangling carbon atoms (Coulson 1960, Menendez et al. 1996a). As a result, re-exposure of such carbons to the atmosphere results in oxygen adsorption and the consequent loss of basic and hydrophobic properties. However, Menendez et al. (1996a, 1996b) recently showed that heat treatment in a hydrogen-atmosphere produces relatively stable carbon atoms at the edges of the graphitic platelets that no longer contain unpaired electrons (Coulson 1960, Menendez 1996a). Consequently, oxygen readsorption on carbons heat-treated in H_2 was greatly reduced compared to carbons heat-treated in N_2. The use of ammonia (NH_3) has also been explored for the preparation of basic activated carbons (Janssen and van Bekkum 1994, Biniak et al. 1997, Mangun et al. 2001). Apart from the removal of oxygen-containing functional groups during heat-treatment in a NH_3 atmosphere, the incorporation of basic nitrogen moieties such as pyridine-type structures can explain the enhanced basicity of ammonia-treated carbons (Janssen and van Bekkum 1995, Biniak et al. 1997, Mangun et al. 2001).

Numerous studies have shown that increases in the oxygen or acidic functional group contents of activated carbons impair the adsorption of organic compounds from aqueous solution (Kaneko et al. 1989, Pendleton et al. 1997, Karanfil and Kilduff 1999, Franz et al. 2000, Considine et al. 2001). Although oxidation of activated carbons enhanced the removal of polar organic compounds from the gaseous phase (Dimotakis et al. 1995, Mangun et al. 1999), Kaneko et al. (1989) showed that the *removal* of acidic functional groups enhanced the adsorption of both relatively polar and relatively nonpolar organic compounds from aqueous solution. This observation is consistent with the preferential adsorption of water on carbon surfaces containing oxygen groups (Kaneko et al. 1989, Pendleton et al. 1997, Karanfil and Kilduff 1999, Franz et al. 2000, Considine et al. 2001).

Water can adsorb by means of hydrogen bonds on oxygen-containing functional groups, and clustering of additional water molecules occurs around water molecules adsorbed at these sites (Dubinin et al. 1955, Puri 1970, Barton et al. 1984, Pan and Jaroniec 1996, Müller et al. 1996, Müller and Gubbins 1998, McCallum et al. 1999, Müller et al. 2000). Such water clusters can prevent pollutant access to hydrophobic regions on the activated carbon surface, reduce the interaction energy between the pollutant and the adsorbent surface, and/or effectively block pollutant access to micropores (Coughlin and Ezra 1968, Mahajan et al. 1980, Kaneko et al. 1989, Pendleton et al. 1997, Franz et al. 2000, Müller et al. 2000). Pendleton et al. (1997) also remarked that organic contaminant adsorption from aqueous solution necessitates the displacement of water from the activated carbon surface. Enthalpy measurements showed that water displacement by MIB became increasingly difficult with an increase in the hydrophilic site concentration on the activated carbon surface. Consequently, water adsorption explains why polar activated carbons exhibit a smaller adsorption capacity for trace organic compounds than less polar activated carbons with similar micropore volumes (Newcombe et al. 1997, Pendleton et al. 1997). Overall, the detrimental effects of water adsorption illustrate that adsorbents with stable hydrophobic surfaces bear the greatest promise for the removal of trace organic contaminants from drinking water sources. Nonetheless, adsorbents need to be sufficiently hydrophilic so that they are wetted by water to be useful for water treatment applications (Davis and Powers 2000).

NOM ADSORPTION

Natural organic matter (NOM) readily adsorbs on activated carbon and interferes with the adsorption of targeted micropollutants. NOM, which is ubiquitous in drinking water sources, consists of (1) degradation products of plant and animal matter, (2) compounds that are biologically or chemically synthesized from these degradation products, and (3) microorganisms and their exudates (e.g., Schnitzer and Khan 1972). The molecular weight of NOM can vary from <500 to > 10^6 Daltons (e.g., Schnitzer and Khan 1972). Aquatic NOM typically consists of compounds with comparatively low molecular weights, and high performance size exclusion chromatography (HPSEC) data of aquatic NOM yielded molecular weight distributions in the range of about 300 to 10,000 Da (Pelekani and Snoeyink 1999, Pelekani et al. 1999, Ebie et al. 2001). For stream fulvic acids, number-averaged molecular weight estimates range from about

600 to 1400 Da (Aiken et al. 1994, Chin et al. 1994) and weight-averaged molecular weight estimates from about 1000 to 2300 Da (Chin et al. 1994). With respect to molecular size, Aiken et al. (1994) observed radii of gyration of 0.7 and 1.1 nm for Suwannee River fulvic and humic acid, respectively.

The above results suggest that NOM can adsorb in the mesopores and large micropores of activated carbons. Numerous researchers have correlated the adsorption of NOM with the pore structure of activated carbons (e.g., Lee et al. 1981, Summers and Roberts 1988b, Ebie et al. 1995, Newcombe et al. 1997, Pelekani and Snoeyink 1999, Ebie et al. 2001). In agreement with the molecular size estimates of Aiken et al. (1994), the results of the latter studies showed that NOM adsorbs in mesopores and large micropores. E.g., Pelekani and Snoeyink (1999) showed that the adsorption of humic substances in a non-fractionated groundwater (MW<3000 Da) was negligible on ACFs that contained pores that were primarily smaller than 0.8 nm in width. In contrast, all size ranges of the same NOM adsorbed on an ACF that contained pores in the entire micropore range. Newcombe et al. (1997) showed that the adsorption capacity for the 500-3000 Da NOM fraction correlated best with the cumulative volume of pores with pore widths ranging from 0.8 to 50 nm, while the adsorption capacity for the >3000 Da NOM fraction correlated well with the mesopore volume. Ebie et al. (1995) showed that the adsorption of non-fractionated humic and fulvic acids with molecular weights between about 500 and 50,000 Da correlated best with the volume of pores with pore widths in the 3 to 10 nm range (i.e., with small mesopores).

Because of the presence of carboxylic acid and phenolic groups in the NOM structure, NOM carries a negative charge at pHs typically encountered in drinking water treatment (Perdue and Lytle 1983). Consequently, electrostatic interactions can play an important role in the adsorption of NOM. Because of the amphoteric nature of activated carbons, pore surfaces can be positively or negatively charged or neutral, depending on the pH of the water being treated as well as on the base material and activation process used to prepare the activated carbon. In aqueous suspensions, acidic activated carbons, i.e., carbons containing carboxylic acid and phenolic groups, carry a negative surface charge at neutral pH while basic carbons carry a positive surface charge. Bjelopavlic et al. (1999) examined the acid/base characteristics of seven commercially available activated carbons and observed that chemically activated wood-based carbons exhibited acidic or neutral surfaces while the remaining steam-activated carbons, which were prepared from coal, lignite, and coconut-shells, exhibited neutral or basic surfaces. If the carbon surface has a positive charge, attractive electrostatic forces arise between the NOM and the carbon surface. Repulsive electrostatic forces can occur in the following situations: (1) activated carbon pore surfaces are negatively charged, (2) adsorbed NOM render activated carbon pore surface negatively charged, and/or (3) adsorbed NOM molecules interact at high surface concentrations (Newcombe 1999).

When considering the adsorption of electrolytes, such as NOM, the nature of the electrostatic interaction also depends on the surface concentration of the electrolyte. At low surface concentrations, electrostatic interactions mainly take place between the surface and the adsorbate (Müller et al. 1985). As the surface concentration increases, however, the results of Müller et al. (1985) suggest that the charge of the adsorbed electrolyte begins to control the surface charge. Morris and Newcombe (1993) observed that the activated carbon surface charge becomes increasingly negative with increasing NOM loading. Consequently, repulsive electrostatic interactions should occur between adsorbed NOM molecules as well as between adsorbed NOM and NOM in solution as NOM accumulates on the activated carbon surface. Furthermore, the greater negative charge of the activated carbon surface may adversely affect the adsorption of nonionic micropollutants from drinking water.

Newcombe and Drikas (1997) studied the electrostatic and non-electrostatic effects on adsorption of NOM by two activated carbons. One carbon was an almost neutral coal-based carbon with the majority of its pores in the micropore range. The other carbon was an acidic wood-based carbon with both micropores and mesopores. At pH 3, electrostatic effects were minimized because acidic functional groups on the adsorbent surface and the NOM were protonated. In that case, the mesoporous wood-based carbon adsorbed more NOM. When NOM loadings were normalized with respect to the pore volume in the secondary micropore and mesopore range, the two carbons performed comparably. As the pH was increased, the coal-based carbon exhibited a larger NOM adsorption capacity when normalized with respect to the accessible pore volume, a result that was explained by more favorable electrostatic interactions between NOM and the less acidic coal-based carbon surface. However, the wood-based carbon exhibited a larger NOM adsorption capacity when normalized with respect to adsorbent mass. Therefore, the adsorbent pore size distribution appears to be the principal factor controlling NOM adsorption while electrostatic effects play a secondary role (Newcombe and Drikas 1997).

Karanfil et al. (1999) studied the adsorption of NOM on surface-modified and as-received carbons. The carbons that underwent chemical modification were a chemically activated wood-based carbon and a thermally activated coal-based carbon. The carbons were acid-washed with 2N HCl to remove ash components and alkaline impurities. Subsequently, the carbons were heat-treated to remove oxygen-containing functionalities. Next, the heat-treated carbons were oxidized at 70°C for either 2 hours or 9 hours to increase surface acidity. Finally, a portion of the oxidized carbons was again heat-treated to remove strongly acidic and other CO_2-evolving groups. NOM adsorption was dependent on raw material type, activation conditions, and surface treatment. For example, NOM uptake decreased upon oxidation of the coal-based carbon. Given that surface oxidation increased the acidity of the carbon, but had a negligible effect on pore structure, repulsive electrostatic interactions between the negatively charged activated carbon surface and the negatively charged NOM may have explained the decreased NOM adsorption upon surface oxidation. Furthermore, preferential water adsorption on polar surface groups may have decreased the accessible pore volume for NOM. Subsequent heat treatment restored some of the NOM adsorption capacity. For the wood-based carbon, however, similar surface treatments had no effect on NOM uptake. Karanfil et al. (1999) suggested that differences in the surface chemistry may have existed among the adsorbents that were not characterized by the methods used in their research.

MICROPOLLUTANT ADSORPTION IN THE PRESENCE OF NOM

It has been widely observed that the adsorption capacity of activated carbons for organic micropollutants is lowered in the presence of NOM (Najm et al. 1991a, Qi et al. 1994, Narbaitz and Benedek 1994, Chen et al. 1997, Newcombe et al. 1997, Knappe et al. 1998, Pelekani and Snoeyink 1999, Ebie et al. 2001). The concentration of NOM in drinking water is about 3 to 6 orders of magnitude larger than the concentration of most micropollutants, and the degree of competition is dependent on the initial concentration of the micropollutant (Najm et al. 1991a, Narbaitz and Benedek 1994, Knappe et al. 1998, Pelekani and Snoeyink 1999). The nature of the

NOM (e.g. chemical composition, adsorbability) also affects the extent of competition (Bernazeau et al. 1996, Müller et al. 1996). Several studies have explored the effects of activated carbon pore structure on the adsorption of micropollutants in the presence of competing organic matter (Newcombe et al. 1997, Pelekani and Snoeyink 2000, Pelekani and Snoeyink 2001, Ebie et al. 2001). However, to date, the effect of activated carbon surface chemistry on the adsorption of micropollutants in the presence of NOM remains largely unexplored.

Newcombe et al. (1997) studied the adsorption of 2-methylisoborneol (MIB) in the presence of NOM. Experiments with NOM fractions showed that the smallest size fraction (MW < 500) resulted in a greater reduction in MIB adsorption than larger size fractions, indicating that NOM molecules in a size range similar to that of the target compound directly compete with the target compound for adsorption sites. This observation is consistent with the findings of other researchers who determined that smaller NOM molecules are adsorbed preferentially on activated carbons because a large percentage of internal surface area of activated carbon is located in the micropore region that cannot be accessed by larger NOM molecules (Summers and Roberts 1988b, Kilduff et al. 1996).

Pelekani and Snoeyink (1999) examined the effect of pore size on the adsorption of atrazine from natural water using a series of ACFs with different activation levels. From these experiments, the two main mechanisms in competitive adsorption were determined to be pore blockage and direct competition, and the dominating mechanism was dependent on the ACF pore size distribution. If pores were large enough to admit the micropollutant but not large enough to admit NOM, then pore blockage was the dominant competition mechanism. However, if pores were large enough to admit both NOM and the micropollutant, then direct competition became the dominant competition mechanism. Furthermore, the data showed that pore blockage by NOM as a result of size exclusion more severely reduced the micropollutant capacity than direct competition between NOM and atrazine for adsorption sites. Pelekani and Snoeyink (1999) therefore suggested that adsorbents with a broader pore size distribution in the micropore range reduce pore blockage effects and thus the negative impact of NOM on micropollutant adsorption. In a study looking at the adsorption of four different micropollutants from natural water by three activated carbons, Ebie et al. (2001) also concluded that the extent of competition was lessened when the pore size distribution was broadened and there was a larger volume percentage of pores greater than 30 Å.

To better understand the role of pore size distribution on competitive adsorption, Pelekani and Snoeyink (2000) performed a series of experiments that examined the adsorption of atrazine in the presence of methylene blue (MB), a dye with a molecular size similar to that of atrazine. Again, ACFs were used to elucidate pore size effects. Because the adsorbates were similar in size, direct competition was observed. By increasing the pore volume and shifting the pore size distribution from the primary micropores (< 8 Å) to the secondary micropores (8-20 Å), the effect of competition on atrazine adsorption was lessened because more adsorption sites were available in the pore size region accessible to both atrazine and MB. Adsorption capacities for atrazine and MB correlated best with pore volumes in the 7.5-10 and 7.5-20 Å range, respectively. Furthermore, atrazine could more easily displace MB in the secondary micropores because the adsorption potential was smaller.

A similar study was performed by Pelekani and Snoeyink (2001) in which MB was replaced with congo red (CR), a dye that is larger in size than atrazine. For an ACF that contained principally primary micropores (<8 Å), both pore blockage and pore constriction occurred, the former lowering the atrazine adsorption capacity and the latter slowing the kinetics of adsorption. Once again, by increasing the pore volume and by shifting the micropore size distribution towards the secondary micropores, the effect of NOM competition was lessened. As the pore volume shifted to smaller end of the secondary micropore region, the CR competition mechanism was primarily one of pore constriction that reduced adsorption kinetics but not adsorption capacity. In the presence of a significant secondary micropore volume (8-20 Å), atrazine capacity reductions as a result of direct site competition between atrazine and CR started to become important while reductions in adsorption kinetics as a result of pore constriction were no longer observed. Pelekani and Snoeyink (2001) also proposed that further activation of ACFs yielded a branched pore structure, in which small micropores branch off from mesopores or larger micropores. In this case, both direct competition in mesopores and large micropores and pore blockage at the entrance of small micropores were observed.

MODELING OF COMPETITIVE ADSORPTION EQUILIBRIA

Over the last few decades, many useful design tools have been developed that require data from bench-scale tests and mathematical models to predict the removal of a target compound in a given natural water by a given activated carbon (e.g., Najm et al. 1991a, Knappe et al. 1998, Kilduff and Wigton 1999, Ebie et al. 2001). These design tools are frequently based on the ideal adsorbed solution theory (IAST) developed by Myers and Prausnitz (1965) for mixed-gas adsorption and extended by Radke and Prausnitz (1972) for multi-solute adsorption from dilute aqueous solutions. The IAST is capable of predicting adsorption equilibria in a multi-solute system provided that single-solute isotherm parameters, such as the Freundlich parameters (K and 1/n) of each solute, as well as the initial solute concentrations (C_0) are known. Underlying assumptions of the IAST include that adsorption sites are equally accessible to all solutes and that adsorption is reversible. While the IAST successfully describes competitive adsorption equilibria in multi-solute systems containing uniquely identifiable compounds such as SOCs, the main difficulty enters when heterogeneous NOM is included as a competing adsorbate.

Kilduff and Wigton (1999) attempted to model the adsorption of TCE in the presence of humic acid. Single-solute TCE adsorption isotherms were modeled using the Langmuir-Freundlich model, and NOM isotherms were modeled by using the Freundlich parameters describing the DOC isotherm for the low-molecular weight humic fraction (< 3000 Daltons). However, the IAST prediction overestimated the solid-phase TCE concentration at high liquid-phase concentrations and underestimated the solid-phase TCE concentration at low liquid-phase concentrations. Kilduff and Wigton (1999) suggested that the inaccurate predictions were caused by the assumption that all sites are equally accessible. The model was therefore adjusted to account for (1) adsorption sites that were accessible to TCE molecules but not to humic molecules and (2) a reduction in effective surface area as a result of NOM adsorption that may have led to pore blockage. These adjustments led to a model that described the data sets well.

Ebie et al. (2001) attempted to model the adsorption of four agricultural chemicals in the presence of NOM. The Freundlich fits of the bulk DOC isotherms were used to model the competing NOM. However, the IAST predictions did not describe the experimental data well. By adding a component to the model to account for the change in capacity due to pore blockage, the SOC isotherms in the presence of NOM could be successfully predicted.

However, pore blockage may not be the only reason why parameters obtained from bulk DOC isotherms do not appropriately describe the competing NOM in the IAST. Studies have shown that the low molecular weight fraction of NOM is preferably adsorbed (Newcombe et al. 1997, Kilduff et al. 1996). Newcombe et al. (1997) proposed that only a very low concentration of low molecular weight (<500 Daltons) NOM is necessary to lower the MIB adsorption capacity via direct competition. Therefore, it is possible that a small fraction of the NOM directly competes with SOCs for adsorption sites, and that the initial concentration and isotherm parameters of this NOM fraction differ greatly from those describing the bulk DOC. Furthermore, current fractionation and analytical techniques for NOM characterization have not been able to identify that fraction of the NOM that competes directly with micropollutants for adsorption sites. For example, a study of three natural waters by Bolto et al. (1999) showed that a DOC fraction always existed that did not absorb light in the UV range. Therefore, a fraction of the low molecular weight NOM may not be detected by UV detectors used to conduct HPSEC.

Another approach for modeling the competing NOM makes use of the adsorption analysis procedure (Sontheimer et al. 1988), in which a dissolved organic carbon (DOC) isotherm is fitted with the IAST using several fictive components. Once the fictive components are known for a given natural water, only the single-solute isotherm parameters for a target compound are required to predict its adsorption equilibrium in that water using the IAST. The adsorption analysis procedure was successfully applied to predict adsorption equilibria of metazachlor in two different natural waters at different initial concentrations (Haist-Gulde 1991, Haist-Gulde et al. 1995). The advantage of the adsorption analysis is that only single-solute isotherm parameters are required to predict adsorption equilibria in a natural water once the DOC isotherm of that water has been described adequately. However, it is not always easy to obtain good DOC isotherms or the adsorption characteristics of the fictive components.

Alternatively, Najm et al. (1991a) proposed that the NOM be modeled as a single component, the equivalent background compound (EBC). This approach also requires one isotherm with the target organic compound in organic-free water and one isotherm with the target organic compound in the natural water. However, substantially less computer programming is required since the competing NOM is modeled as one compound. Several studies found this approach to be successful (Najm et al. 1991a, Knappe et al. 1998, Pelekani and Snoeyink 1999, Graham et al. 2000). Graham et al. (2000) used the EBC approach to model the adsorption of MIB and geosmin from four natural waters. When the K, 1/n, and molecular weight of the EBC were kept constant and the initial concentration was allowed to vary for each water, they were able to model all four waters using a single EBC. Furthermore, for three of the waters, the initial concentration of the EBC was approximately 0.45% of the total initial TOC concentration. This result supports the theory that only a small portion of the NOM directly competes with the target compound for adsorption sites.

For trace organic contaminant adsorption in the presence of NOM, Knappe et al. (1998) observed a direct proportionality between the PAC adsorption capacity and the initial micropollutant concentration at a given carbon dose. Using the EBC modeling approach, Knappe et al. (1998)

showed that this observation was consistent with the IAST, provided that the SOC was present in trace amounts. Consequently, mathematical models are not needed to predict the initial concentration dependence of micropollutant isotherms in natural water, provided that a micropollutant isotherm experiment was conducted at a sufficiently low initial concentration in the natural water of interest.

SUMMARY

The above review of the literature illustrates that numerous studies have evaluated the effects of activated carbon surface chemistry and pore structure on the adsorption of nonionic trace organic compounds and NOM. Table 2.1 summarizes the most important factors that control the adsorption of both compound categories. In addition, Table 2.1 illustrates that important gains have been made in terms of understanding the effects of activated carbon pore structure on the adsorption of trace organic compounds in the presence of NOM. However, knowledge with respect to the effects of activated carbon surface chemistry on trace organic compound adsorption in the presence of NOM is still lacking to date. To develop activated carbon selection criteria, it is therefore important to obtain a clearer understanding of the effects of activated carbon surface chemistry on the adsorption of trace organic compounds from natural water.

Table 2.1 Effects of activated carbon surface chemistry and pore structure on adsorption

Adsorbent characteristic	Nonionic trace organic compounds	NOM	Nonionic trace organic compounds in the presence of NOM
Pore size distribution	Require small micropores	Requires large micropores and small mesopores	pore blockage by NOM for carbons that contain only small micropores – detrimental effect on target compound adsorption direct competition between NOM and target compound for carbons that contain larger micropores and mesopores – less detrimental on target compound adsorption
Surface chemistry	Require hydrophobic surfaces to prevent detrimental effects of water adsorption	Little effect	little information
Surface charge	Little effect	Enhanced adsorption on basic carbons because of electrostatic attraction	little information

CHAPTER 3
MATERIALS AND METHODS

SOLVENTS

Isotherm experiments were conducted in ultrapure laboratory water (tap water treated by reverse osmosis, ion exchange, and granular activated carbon adsorption, resistance ≥ 14.85 MΩ/cm). Ultrapure water was amended with a 1 mM phosphate buffer (0.5 mM $Na_2HPO_4 \bullet H_2O$ and 0.5 mM NaH_2PO_4) to maintain a pH of 7.2. Furthermore, 100 mg/L sodium azide were added to eliminate aerobic biological activity. To evaluate solvent effects, adsorption experiments were also conducted in cyclohexane (HPLC-grade, Sigma-Aldrich, St. Louis, MO), a nonpolar solvent.

NOM adsorption and micropollutant adsorption in the presence of NOM were evaluated with Sacramento-San Joaquin Delta Water (SJDW). The water was sampled at the intake of the Concord, CA, water treatment plant (Contra Costa Water District), filled into 55-gallon stainless steel drums, and immediately shipped by 3-day freight service to the Environmental Engineering Laboratory at NCSU. Upon arrival, the water was preserved with 100 mg/L sodium azide. Prior to use in isotherm experiments, the water was vacuum filtered through a 0.45-μm nylon membrane filter (Magna-R, MSI, Westboro, MA) that was placed in a 47 mm glass microanalysis filter holder (Fisher Scientific, Pittsburgh, PA). The DOC of SJDW was 4.0 mg/L, the pH was 7.9, the total alkalinity was 53 mg/L as $CaCO_3$, and the total hardness was 72 mg/L as $CaCO_3$.

ADSORBATES

Trichloroethene (= trichloroethylene or TCE) and methyl tertiary-butyl ether (MTBE), two common drinking water contaminants, served as probe molecules for the assessment of adsorbate polarity effects. TCE is a relatively hydrophobic adsorbate [aqueous solubility at 25°C = 1,198 mg/L (Schwarzenbach et al. 1993)] while MTBE is relatively hydrophilic [aqueous solubility ≈ 50,000 mg/L (Nichols et al. 2000)]. In addition, TCE and MTBE differ in molecular size and shape (Figure 3.1); TCE is a planar molecule while MTBE approximates a tetrahedron. Kinetic diameters for TCE and MTBE are 5.6 Å (Chintawar and Greene 1997) and 6.2 Å (Sano et al. 1995), respectively.

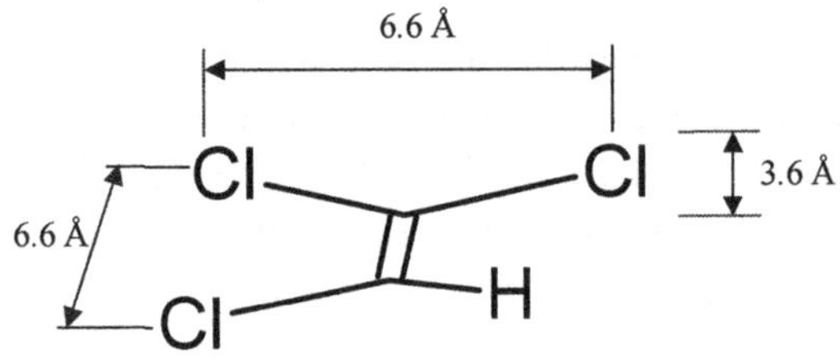

Trichloroethene

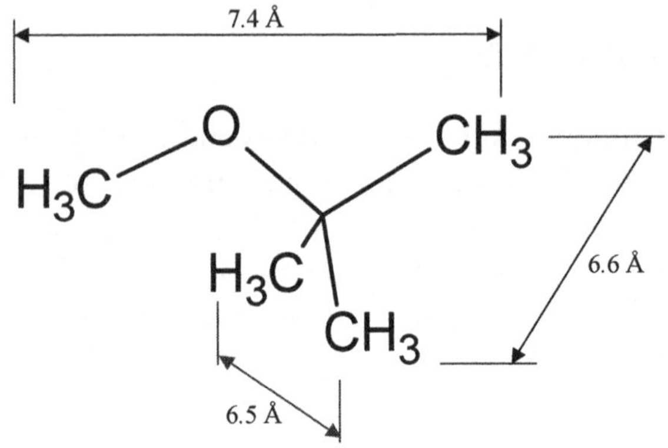

Methyl tertiary-butyl ether

Figure 3.1 Molecular structures and dimensions for trichloroethene and methyl tertiary-butyl ether. Molecular dimensions were calculated from the Cambridge Structural Database and the van der Waals radii for methyl groups and chlorine [2.0 and 1.8 Å, respectively (Data from Pauling 1960)].

For single-solute isotherms, MTBE stock solutions were prepared in ultrapure water at a concentration of approximately 1,500 mg/L, and TCE stock solutions were prepared in methanol at a concentration of about 3,400 mg/L. For isotherms in natural water, MTBE and TCE stock solutions were prepared at a concentration of about 265 mg/L in ultrapure water and methanol, respectively. TCE stock solution concentrations assured that methanol concentrations in isotherm tests remained below 0.1%, a level at which no co-solvent effects were observed in prior studies (Kilduff et al. 1998).

ADSORBENTS

Three commercially available ACF samples were obtained from Nippon Kynol (ACF-10, ACF-15, and ACF-20). The phenolic-resin fibers were pyrolyzed and subsequently oxidized in a steam and CO_2 atmosphere to obtain three activation levels (ACF-10 lowest, ACF-20 highest). ACFs had the appearance of charred cotton balls and were cut with scissors into 0.5-cm (0.2-inch) pieces prior to surface chemistry modifications. All ACF samples were acid-washed by soaking the ACF overnight in 2 N HCl and subsequently boiling the ACF for 1 hour in 2 N HCl. The samples were then placed in a glass column and rinsed with ultrapure water until no chloride was detected by $AgNO_3$ addition to the rinse water. Acid-washed ACF samples were dried in an oven at 130°C and served as the base material for subsequent surface chemistry modifications.

Acid-washed ACFs were oxidized by contacting 5 g of ACF with 200 mL of 30% H_2O_2 for 16 hours at room temperature. Following oxidation, the ACF was dried in an oven at 130°C. Heat treatment in a hydrogen atmosphere followed a procedure similar to that described by Menendez et al. (1996b). A 1.5-g sample of acid-washed ACF was mixed with 1 g of platinum-impregnated GAC (Alfa Aesar, Ward Hill, MA), and the mixture was placed into a 3-zone tube furnace (Lindberg/Blue M, Asheville, NC) equipped with an alumina tube (7.62-cm O.D., Vesuvius McDanel, Beaver Falls, PA). The furnace temperature was raised to 900°C, held at this temperature for 2 hours and then cooled to 500°C, all in a N_2 atmosphere. At 500°C, the gas supply was switched to H_2 for 3 hours. Finally, ACF samples were cooled to room temperature in the tube furnace under N_2. Upon removal of the ACF/GAC mixture from the furnace, the GAC was removed from the ACF with tweezers. For ammonia-treatments, a 1.5-g sample of acid-washed ACF was placed into the tube furnace without catalyst. The furnace temperature was raised to 900°C, held at this temperature for 2 hours and then cooled to 700°C, all in a N_2 atmosphere. At 700°C, the gas supply was switched to NH_3 for 1 hour. Finally, ACF samples were cooled to room temperature in the tube furnace under N_2.

The furnace setup for hydrogen and ammonia treatments is depicted in Figure 3.2. All gases were ultrahigh purity (UHP) or research grade, and gas streams passed through oxygen, carbon dioxide, and water traps prior to entering the reaction tube. All gas flow rates were 200 mL/min. Tube ends were sealed with stainless steel plates, PTFE-lined gaskets, and high temperature grease to prevent leakage of H_2 and NH_3 or intrusion of atmospheric oxygen, CO_2, and water. Upon heat treatment, samples were stored in a desiccator at room temperature. To facilitate comparisons among modified ACFs, the abbreviations shown in Table 3.1 will be used throughout this report.

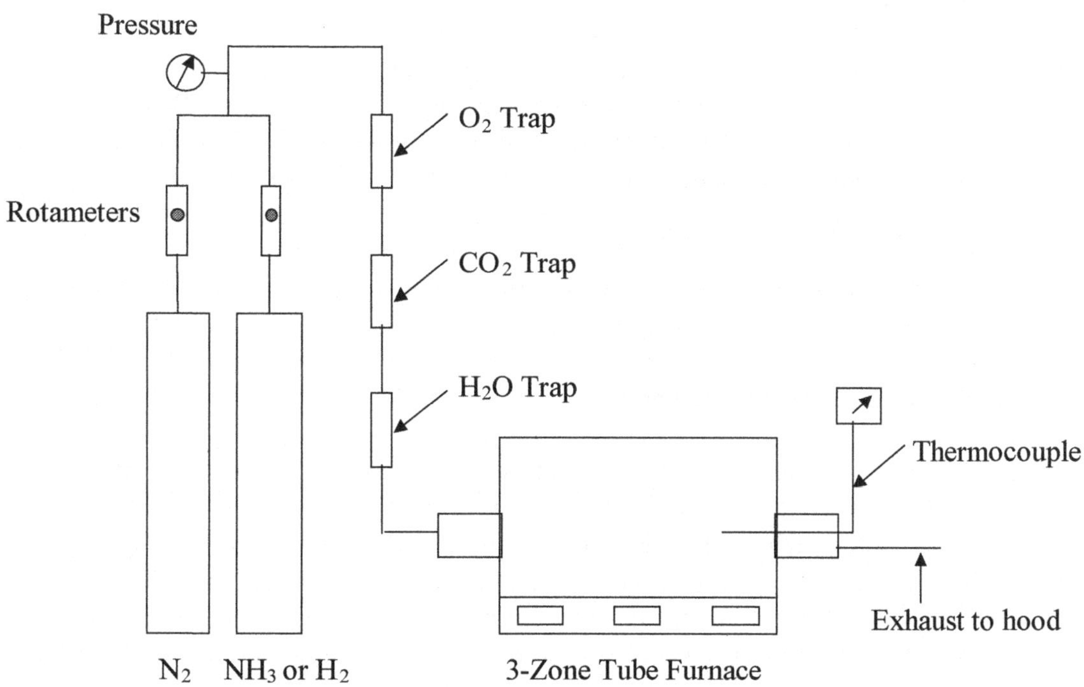

Figure 3.2 Tube furnace setup

Table 3.1 Nomenclature for ACF matrix

Treatment	Abbreviation		
	ACF-10	ACF-15	ACF-20
Acid-washed	AW10	AW15	AW20
Oxidized after acid-washing	OAW10	OAW15	OAW20
Hydrogen-treated after acid-washing	HAW10	HAW 15	HAW 20
Ammonia-treated after acid-washing	AAW10	AAW15	AAW20

To verify whether correlations developed for the ACF matrix are also valid for commercially available granular activated carbons (GACs), tests were conducted with bituminous coal-based F-600 GAC (Calgon Carbon Corporation, Pittsburgh, PA), coconut shell-based G-219 GAC (PicaUSA, Columbus, OH), and chemically activated wood-based Picazine GAC (PicaUSA, Columbus, OH). To enhance adsorption rates, GACs were pulverized with a mortar and pestle until 95% by mass passed a 45-µm sieve (325 U.S. mesh). Upon sieving, the portion remaining on the sieve was recombined with the portion that passed through the sieve to prevent bias as a result of any physical and/or chemical differences between the two fractions. The pulverized GAC was dried at 130°C and stored in a desiccator.

ADSORBENT CHARACTERIZATION

BET surface areas, micropore volumes and pore size distributions were determined from N_2 isotherm data collected at 77 K (Autosorb-1-MP, Quantachrome Corporation, Boynton Beach, FL). Prior to analysis, ACF samples were outgassed overnight at 473 K. BET surface areas were determined from 11-point adsorption isotherms that were completed with a 0.1-g sample in the 0.01-0.1 relative pressure range. Subsequently, N_2 isotherms were obtained over a relative pressure range from 10^{-6} to 1 with 0.01-g sample sizes. The ACF sample mass for the latter N_2 isotherm was adjusted such that the BET surface area matched the value obtained with the larger sample mass. The micropore volumes and pore size distributions were subsequently calculated from the N_2 adsorption data using the density functional theory (Vulcan kernel, PC software version 1.19, Quantachrome, Boynton Beach, FL).

The point of zero charge for activated carbon surfaces was assessed by mass titrations as described by Noh and Schwarz (1989) and Barton et al. (1997). Using carbon doses up to 5% (wt/wt), mass titrations were performed in 25-mL crimp-cap culture tubes containing 20 mL of 0.1 N NaCl solution. To eliminate interferences from dissolved and atmospheric CO_2, the NaCl solution was sparged with N_2 for four hours, and culture tubes were filled and capped in an anaerobic hood. Changes in solution pH as a result of the activated carbon additions were measured in an anaerobic hood after tumbling suspensions for 48 hours. The pH_{PZC} was taken as the average plateau value that was obtained for the largest carbon doses.

Carbon, hydrogen, and nitrogen contents were determined with a CHN Elemental Analyzer (Perkin-Elmer Corp., Norwach, CT). The phosphorous content of Picazine GAC was determined with an ICP Optical Emission Spectrometer (Perkin-Elmer Corp., Norwach, CT). Prior to analysis, the samples were dried overnight at 105°C and cooled in a desiccator. Oxygen contents were measured according to ASTM D5622 (Huffman Laboratories, Golden, CO). Ash contents of GACs were determined by combusting about 5 g of adsorbent in a muffle furnace at 550°C for 2 hours.

For the Infrared Spectroscopy (IR) analysis of ACF samples, the attenuated total reflectance (ATR) technique was employed. Surface functional groups of ACF samples were analyzed by Fourier-Transform IR (FTIR) using an FTS 6000 Spectrometer (Bio-Rad Laboratories, Cambridge, MA) equipped with an MCD detector. Spectra were obtained using a multi-point reflection ATR attachment (Spectra-Tech, Shelton, CT) with a germanium IRE crystal. Germanium was selected because it has a high refractive index (n =4.0), which ensured total reflection of the incident light when analyzing carbons (n=3.3). Samples were placed on the IRE crystal in the sample chamber, which was purged for 20 to 30 minutes with dry air. Each spectrum was obtained at a resolution of 4 cm^{-1} and with 128 co-additions. All IR spectra were background-corrected and baseline-corrected.

ISOTHERMS

For isotherms in ultrapure water, adsorbent doses between 10 and 7500 mg/L were utilized. Adsorbents were placed into brown glass bottles that were subsequently filled to the neck with the amended ultrapure water. TCE or MTBE stock solutions were added with a constant rate syringe (CR-700, Hamilton, Reno, NV) to yield an initial concentration of about 1000 μg/L (some MTBE isotherm experiments were conducted at an initial concentration of approximately 100 μg/L to reduce the amount of carbon required to complete the tests). Once the micropollutant was added to a bottle, it was topped off immediately with ultrapure water to create headspace-free conditions and capped using PTFE-faced silicon septa and open-top closures. A mixing time of 2 weeks in a rotary tumbler was sufficient to reach adsorption equilibrium, as established in screening tests. Micropollutant losses were not observed in triplicate blanks containing no carbon over that time period. Upon equilibration, adsorbents were separated from the liquid by sedimentation (ACFs) or filtration through 0.45 μm filters (pulverized GACs). Aqueous samples were analyzed for MTBE and TCE using a purge and trap concentrator (Tekmar 3100, Cincinnati, OH) followed by a gas chromatograph (Shimadzu 14a, Columbia, MD) with 30 m column (J&W Scientific DB-VRX, I.D. 0.45 mm, liner thickness 2.55 μm, Folsom, CA) and a flame ionization detector (FID). Samples (5 mL) were purged with nitrogen gas at 36°C for 12 min. The analytes were trapped on a Vocarb 3000 trap (Supelco, Bellefonte, PA) and desorbed for 2 min at 255°C. The GC oven temperature was maintained at 40°C for 3 min, was increased at 20°C/min and held at 90°C for 2.5 minutes, then increased at 20°C/min and held at 140°C for 1 minute, and then increased at 40°C/min and held at 240°C for 1 min. The MTBE and TCE concentrations were quantified using external standards.

For micropollutant isotherms in natural water, adsorbent doses between 2 and 5500 mg/L were utilized. Adsorbents were placed into brown glass bottles that were subsequently filled to the neck with SJDW. Immediately after the addition of SJDW to a given isotherm bottle, TCE or

MTBE stock solution was added with a constant rate syringe (CR-700, Hamilton, Reno, NV) to yield an initial concentration of approximately 100 μg/L. Once the micropollutant was added, the bottle was topped off immediately with SJDW to create headspace-free conditions and capped using PTFE-faced silicon septa and open-top closures. A mixing time of 4 weeks in a rotary tumbler was sufficient to reach adsorption equilibrium, and micropollutant losses were not observed in blanks containing no carbon over that time period. Upon equilibration, adsorbents were separated from the liquid by sedimentation (ACFs) or filtration through 0.45 μm filters (pulverized GACs), and remaining liquid phase concentrations were measured by GC as described above.

For NOM isotherms, adsorbent doses between 10 and 2000 mg/L were utilized. Adsorbents were transferred into 4 oz, 8 oz, or 16 oz amber glass bottles depending on the carbon dose. The bottles were subsequently filled with 100 mL (4 oz bottles), 200 mL (8 oz bottles), or 400 mL (16 oz bottles) of SJDW and capped. A mixing time of 2 weeks in a rotary tumbler was sufficient to reach adsorption equilibrium as determined by kinetic tests. Upon equilibration, samples were filtered through 0.22 μm nylon membrane filters (Magna-R, MSI, Westboro, MA) that were placed in a 25 mm stainless steel syringe filter holder (Fisher, Pittsburgh, PA). The filters were soaked overnight in organic-free water prior to use. DOC levels were quantified using a TOC-5000A TOC analyzer (Shimadzu, Columbia, MD). All samples were acidified with 1 N HCl and sparged with air for 4 minutes to measure non-purgeable organic carbon (NPOC). UV absorbance measurements were conducted on filtered samples prior to acidification using a UV/vis spectrophotometer (Unicam UV 1, Spectronic Unicam, Cambridge, United Kingdom). Samples were analyzed in a 1-cm quartz cell at a wavelength of 280 nm, a wavelength at which sodium azide did not interfere with UV absorbance measurements.

For adsorption tests in cyclohexane, ACF doses of 10 and 15 g/L were tested in 40-mL EPA vials. TCE or MTBE were spiked into cyclohexane to yield an initial concentration of 100 mg/L. Thirty-mL aliquots of this solution were pipetted into the vials, which were capped with PTFE-faced silicon septa and open-top closures. The samples were tumbled for one week, and remaining liquid phase concentrations were analyzed by direct injection into a gas chromatograph (HP 5890, Palo Alto, CA) equipped with 75-m column (J&W Scientific DB-VRX, I.D. 0.45 mm, liner thickness 2.55 μm, Folsom, CA) and a flame ionization detector (FID). The GC oven temperature was maintained at 40°C for 5 min and was increased to 200°C at 20°C/min. The MTBE and TCE concentrations were quantified using external standards. Blanks containing no carbon showed no adsorbate losses over the equilibration time.

MATHEMATICAL MODELING OF COMPETITIVE ADSORPTION EQUILIBRIA

TCE and MTBE adsorption in the presence of competing NOM was modeled using the ideal adsorbed solution theory (IAST), and the competing NOM fraction was described by a single, hypothetical compound, the equivalent background compound (EBC) (Najm et al. 1991a). Knowing the sinlge-solute Freundlich isotherm parameters for the micropollutants, the EBC parameters (K_{EBC}, $1/n_{EBC}$, and $C_{o,\ EBC}$) were determined by minimizing the error between the experimental micropollutant isotherm data in the presence of NOM and the IAST model output. To better compare the competitive effect of the NOM on the performance of the tested adsorbents, the EBC parameters for each micropollutant isotherm in SJDW were obtained by varying K and C_0 and by holding the 1/n value constant at 0.55. For a given data set, feasible solutions were obtained for Freundlich exponent values in the range of about 0.45 to 0.63, and the chosen Freundlich exponent of 0.55 represents an average value of this range. This 1/n value is similar to Freundlich exponents reported in prior studies that employed the EBC or other fictive component methods (e.g., Knappe et al. 1998, Graham et al. 2000, Newcombe et al. 2002). Apart from the isotherm curvature at very low carbon doses, K values of about 450 (μmol/g)(L/μmol)$^{1/n}$ [or 10 (mg/g)(L/μg)$^{1/n}$] and greater had no effect on the IAST model output (Figure 3.3). Observing a K-value of about 450 (μmol/g)(L/μmol)$^{1/n}$ or greater suggests the presence of a highly adsorbable NOM fraction that competed with the micropollutants for adsorption sites. Nonetheless, a K-value of this magnitude appears to be reasonable, especially when considering that K values on the order of 10,000 (μmol/g)(L/μmol)$^{1/n}$ have been reported for strongly adsorbing micropollutants such as heptachlor (Snoeyink and Summers 1999). Consequently, the initial EBC concentration was the principal parameter affecting the micropollutant isotherm fits in the presence of NOM.

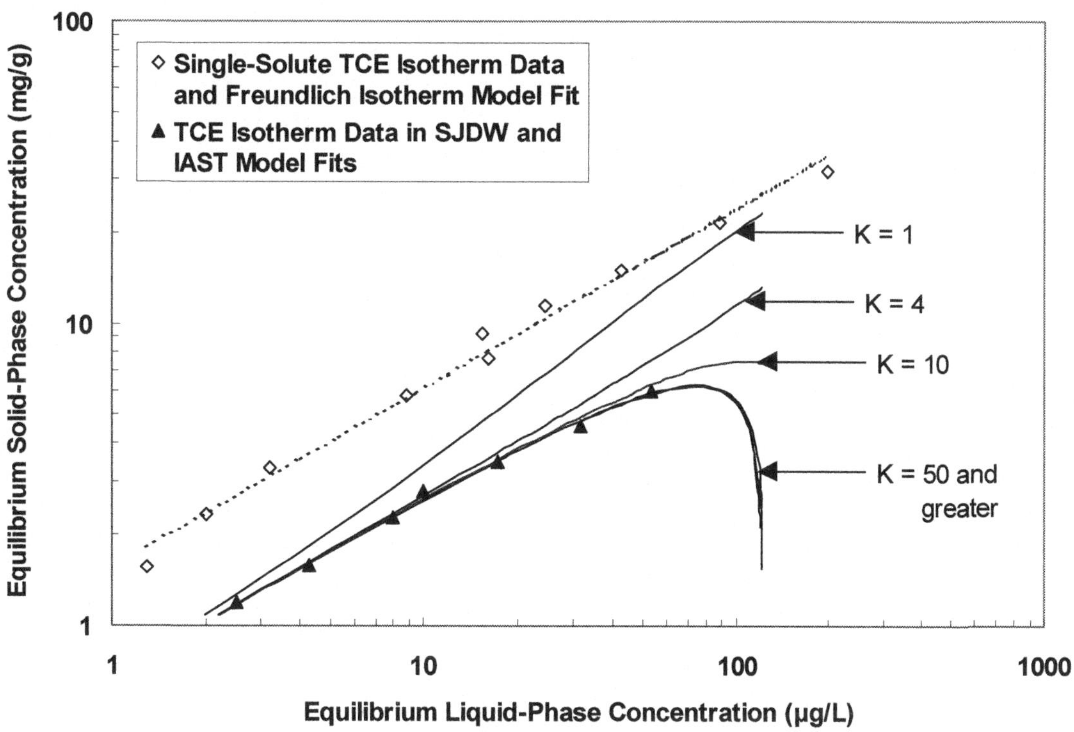

Figure 3.3 Dependence of IAST model fits on K_{EBC}. Experimental data are for TCE adsorption on AW20, and initial TCE concentration in SJDW was 122 μg/L. Units of K_{EBC} are (mg/g)(L/μg)$^{1/n}$, $1/n_{EBC}$ was 0.55, $C_{o,EBC}$ was 5100 μg/L, and MW_{EBC} was 1000 g/mol.

CHAPTER 4
RESULTS AND DISCUSSION

The objectives of this study were (1) to determine the effects of activated carbon pore structure and surface chemistry on the adsorption of organic micropollutants from drinking water and (2) to develop criteria that aid utilities in selecting the most appropriate carbons for a given application. To accomplish these objectives, a matrix of twelve activated carbon fibers (ACFs) with three activation levels and four surface chemistry levels was prepared. To test whether the trends established with the ACFs were also valid for granular activated carbons (GACs), three commercially available GACs were also tested in pulverized form. Adsorbents were first characterized by determining their surface area, pore structure, elemental composition, acid/base characteristics, and surface chemistry. Subsequently, single-solute isotherm experiments were conducted with two adsorbate probes: methyl tertiary-butyl ether (MTBE), a relatively hydrophilic adsorbate, and trichloroethene (TCE), a relatively hydrophobic adsorbate. Furthermore, the adsorption of natural organic matter (NOM) was evaluated for each adsorbent, and the effects of NOM on micropollutant adsorption capacities were evaluated in simultaneous loading experiments.

ADSORBENT CHARACTERIZATION

Physical Characteristics

Table 4.1 summarizes BET surface areas as well as micropore and mesopore volumes of the adsorbents. The results illustrate that the BET surface areas of the ACFs increased with increasing level of activation; i.e., from ACF 10 to ACF 20. Increases in BET surface area were primarily attributable to increases in micropore volume although the mesopore volumes of the ACFs also increased somewhat with increasing activation level. The BET surface area data in Table 4.1 closely agree with results of prior studies that employed phenolic resin-based ACFs with the same designations (e.g., Mangun et al. 1999, 2001, Pelekani and Snoeyink 2000, 2001). Furthermore, micropore volume data calculated from the density functional theory (Table 4.1) were similar to those obtained from the Dubinin-Radushkevich equation in prior studies (e.g., Mangun et al. 1999, 2001, Pelekani and Snoeyink 2000, 2001). The results in Table 4.1 further illustrate that two of the tested GACs, F600 and G219, were principally microporous. While the micropore volume of Picazine GAC was similar to that of the other GACs, its mesopore volume was largest among the adsorbents evaluated in this study.

Table 4.1 BET surface areas and pore volumes of ACFs and pulverized GACs. Micropore and mesopore volumes were obtained from density functional theory calculations.

Sample	BET surface area (m2/g)	Micropore volume (cm3/g)	Mesopore volume (cm^3/g)
AW10	760	0.387	0.075
OAW10	750	0.340	0.089
HAW10	830	0.414	0.074
AAW10	720	0.371	0.061
AW15	1480	0.567	0.099
OAW15	1350	0.439	0.154
HAW15	1530	0.593	0.085
AAW15	1510	0.573	0.098
AW20	1670	0.594	0.120
OAW20	1530	0.504	0.144
HAW20	1710	0.622	0.096
AAW20	1700	0.599	0.130
F600	1290	0.353	0.085
G219	1460	0.438	0.118
Picazine	1680	0.377	0.252

Figure 4.1 depicts micropore size distributions for acid-washed ACFs. In Figure 4.1, the ordinate value associated with each data point represents the pore volume in a 1-Å pore width increment; e.g., the data point for AW10 at a pore width of 6.5 Å illustrates that a pore volume of 0.038 cm^3/g was present in the 6 to 7 Å pore width increment. Micropores of AW10 were concentrated in the 7 to 11 Å width range while those of AW15 and 20 were concentrated in the 9 to 13 Å width range. The micropore size distributions obtained in the current study are in close agreement with prior results for as-received ACFs with the same designations (Daley et al. 1996). As shown in Table 4.1, the chemical modification techniques changed BET surface areas by less than 10% compared to the corresponding acid-washed ACFs. These changes were relatively small given that the BET surface areas of the acid-washed ACFs varied by more than a factor of two. Duplicate BET surface area determinations for two adsorbents showed that coefficients of variation [CV = (standard deviation/mean)*100%] were less than 1%, however, suggesting that the changes in BET surface area upon surface chemistry modifications were generally larger than the experimental error associated with the measurement. Oxidative treatments reduced the micropore volumes of the activated carbons by about 12 to 23% while heat treatments in H_2 or NH_3 atmospheres had little effect on micropore volumes (4% decrease to 7% increase). Duplicate micropore volume determinations for two adsorbents showed that CVs

were less than 3%; therefore, the decreases in micropore volume resulting from oxidative treatments exceeded the experimental error associated with the measurement. With the exception of OAW15, mesopore volume changes were in the ±20% range and were relatively small (in terms of cm^3/g) given the microporous nature of the ACFs.

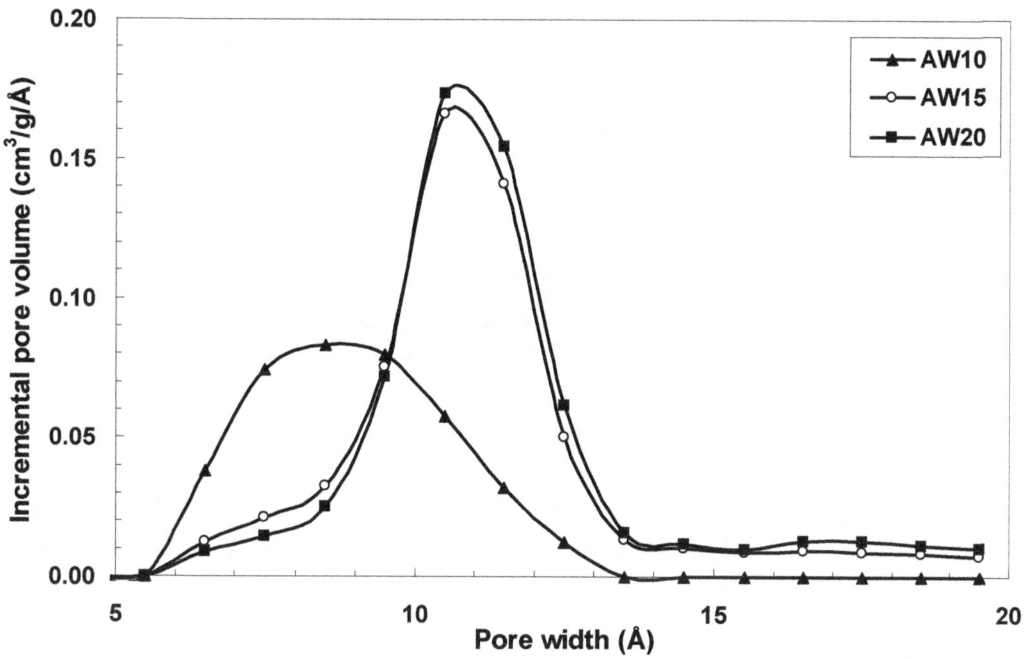

Figure 4.1 Micropore size distributions of acid-washed ACF samples

Although surface chemistry modifications had small effects on BET surface areas and pore volumes, they had no noticeable effects on pore size distributions as illustrated in Figure 4.2. Therefore, the ACF matrix permitted the evaluation of surface chemistry effects on TCE and MTBE adsorption in the absence of pore size effects. For the commercially available GACs, Figure 4.3 shows that both F600 and G219 exhibited pore size distributions that were broader and less uniform than those of the ACFs. Micropores of F600 were concentrated in the 7 to 12 Å width range while those of G219 were concentrated in the 7 to 13 Å width range. The micropores of Picazine were largest among the commercially available GACs and were concentrated in the 9 to 15 Å width range.

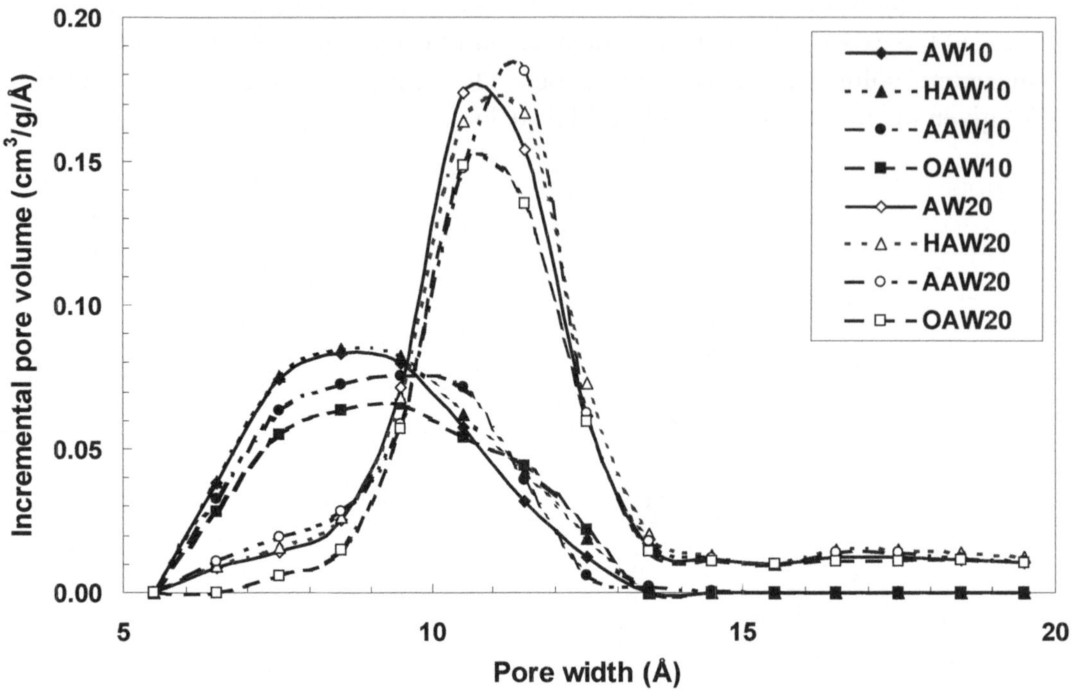

Figure 4.2 Effects of surface chemistry modifications on micropore size distributions of ACFs 10 and 20. Trends for ACF 15 were similar to those observed for ACF 20 (data for ACF 15 not shown for clarity of figure).

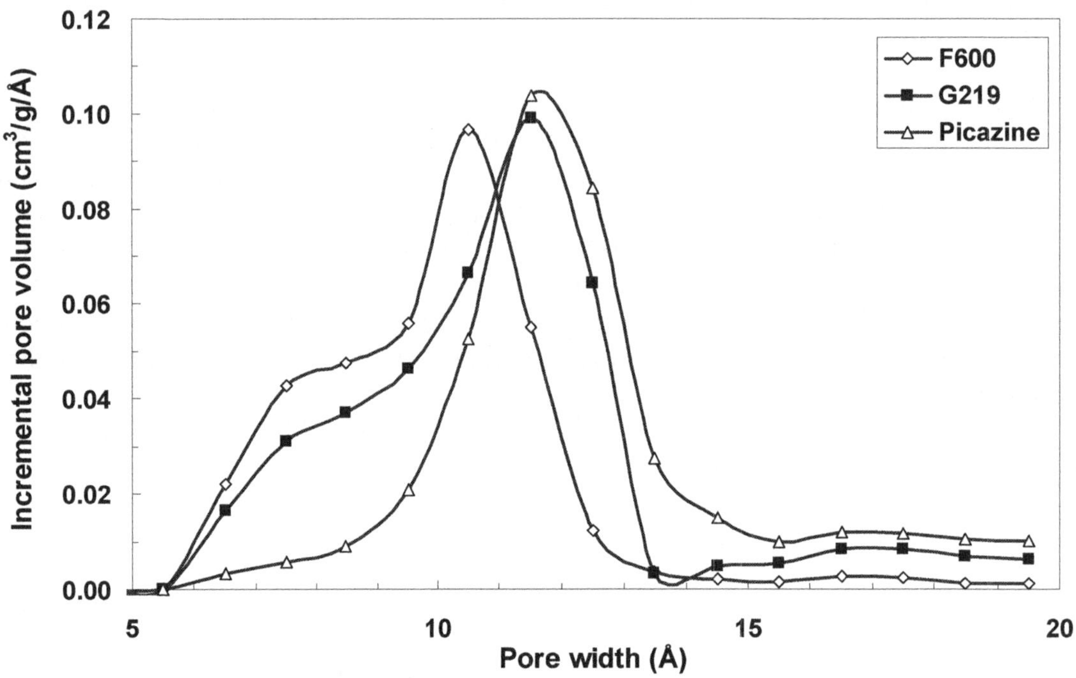

Figure 4.3 Micropore size distributions of pulverized GAC samples

Chemical Characteristics

Elemental analysis and pH_{PZC} data for the ACF matrix are summarized in Table 4.2. The sum of the C, H, N, and O contents represented approximately 94-99% of the total adsorbent mass, confirming that the ash content of the phenolic resin-based ACFs was small. Table 4.2 further illustrates that oxidation of acid-washed ACFs increased the oxygen content to about 10% while heat treatments of acid-washed ACFs in hydrogen and ammonia atmospheres decreased oxygen contents to about 2% or less. Using H_2O_2 to oxidize ACF 15, Mangun et al. (1999) also obtained an oxygen content of about 10%, and XPS analyses showed that this treatment increased the surface concentrations of carboxylic acid, phenolic hydroxyl, and carbonyl groups. As observed in many prior studies (e.g., Barton et al. 1997), the incorporation of oxygen-containing functionalities as a result of surface oxidation rendered the adsorbents more acidic while the removal of oxygen-containing functionalities through heat-treatment rendered the adsorbents more basic. Ammonia-treatment increased nitrogen contents of the ACFs to about 1.5 to 1.9%. Consistent with the incorporation of basic pyridine-type structures, which was expected based on the treatment conditions (Janssen and van Bekkum 1994, Biniak et al. 1997, Janssen and van Bekkum 1995), the pH_{PZC} data illustrate that the ammonia-treatment yielded the most basic ACFs of the adsorbent matrix. The results of the elemental analyses and mass titrations indicate that a given chemical treatment yielded ACFs with similar heteroatom content and surface acidity/basicity. Thus, the ACF matrix also permitted the study of pore structure effects on trace organic compound adsorption with little interference from surface chemistry effects.

Table 4.2 Elemental analysis and pH_{PZC} results for ACF matrix[a]

ACF	C (Wt. %)	H (Wt. %)	N (Wt. %)	O (Wt. %)	Total (%)	pH_{PZC}
AW10	90.18 ± 0.31	0.34 ± 0.02	0.17 ± 0.02	4.57	95.26	4.8
AW15	92.73 ± 0.29	0.40 ± 0.17	0.18 ± 0.01	3.62	96.93	5.4
AW20	92.33 ± 0.15	0.27 ± 0.01	0.15 ± 0.01	3.58	96.33	4.4
OAW10	82.21 ± 1.17	0.54 ± 0.04	0.14 ± 0.01	10.78	93.68	3.0
OAW15	82.37 ± 0.84	0.36 ± 0.02	0.19 ± 0.02	11.34 ± 0.54	94.25	2.9
OAW20	86.30 ± 0.48	0.69	0.09	9.58 ± 0.06	96.66	3.2
HAW10	93.14 ± 0.92	0.36 ± 0.17	0.11 ± 0.03	2.32 ± 0.19	95.92	6.4
HAW15	96.00 ± 0.60	0.27 ± 0.03	0.09 ± 0.01	1.39	97.74	6.5
HAW20	95.68 ± 0.23	0.24 ± 0.02	0.09 ± 0.02	1.66	97.68	6.7
AAW10	92.22 ± 0.81	0.38 ± 0.08	1.80 ± 0.63	2.08 ± 0.17	96.47	8.7
AAW15	95.26 ± 0.55	0.73 ± 0.06	1.89 ± 0.23	1.45 ± 0.16	99.33	8.7
AAW20	93.50 ± 0.53	0.29 ± 0.04	1.52 ± 0.66	1.23 ± 0.45	96.54	8.7

[a] where applicable, values represent means ± one standard deviation

Table 4.3 summarizes the elemental composition, ash content, and pH_{PZC} data for pulverized GACs. The results showed that the low-ash GACs exhibited a wide range of oxygen contents. F600 and G219, the GACs with the lower oxygen contents, exhibited a basic character while Picazine with an oxygen content of nearly 16% was strongly acidic. Because Picazine is chemically activated using phosphoric acid, one could argue that a fraction of the measured oxygen content was associated with phosphates instead of the carbon surface. A phosphorous content of approximately 1% (wt/wt) was measured for Picazine, suggesting that no more than approximately 2% (wt/wt) oxygen was associated with phosphates. Consequently, the largest fraction of the measured oxygen content of Picazine (Table 4.3) was associated with the carbon surface, which explains its acidic character.

Table 4.3 Elemental analysis, ash content, and pH_{PZC} results for pulverized GACs[a]

Carbon	C (Wt. %)	H (Wt. %)	N (Wt. %)	O (Wt. %)	Ash (Wt. %)	Total (%)	pH_{PZC}
F600	92.50 ± 1.68	0.61 ± 0.07	0.41 ± 0.01	2.60	1.84 ± 0.01	97.96	8.3
G219	90.61 ± 0.45	0.57	0.13	4.90	2.82 ± 0.04	99.03	9.7
Picazine	76.92 ± 0.53	0.40 ± 0.03	0.03 ± 0.03	15.88	3.37 ± 0.11	96.59	2.1

[a] where applicable, values represent means ± one standard deviation

To further evaluate the effects of chemical modifications on ACF surface chemistry, IR spectra were collected for all members of the ACF matrix (Figures 4.4 to 4.6), and functional group assignments are summarized in Table 4.4. In addition, Figure 4.7 depicts IR spectra for the three GAC samples evaluated in this study.

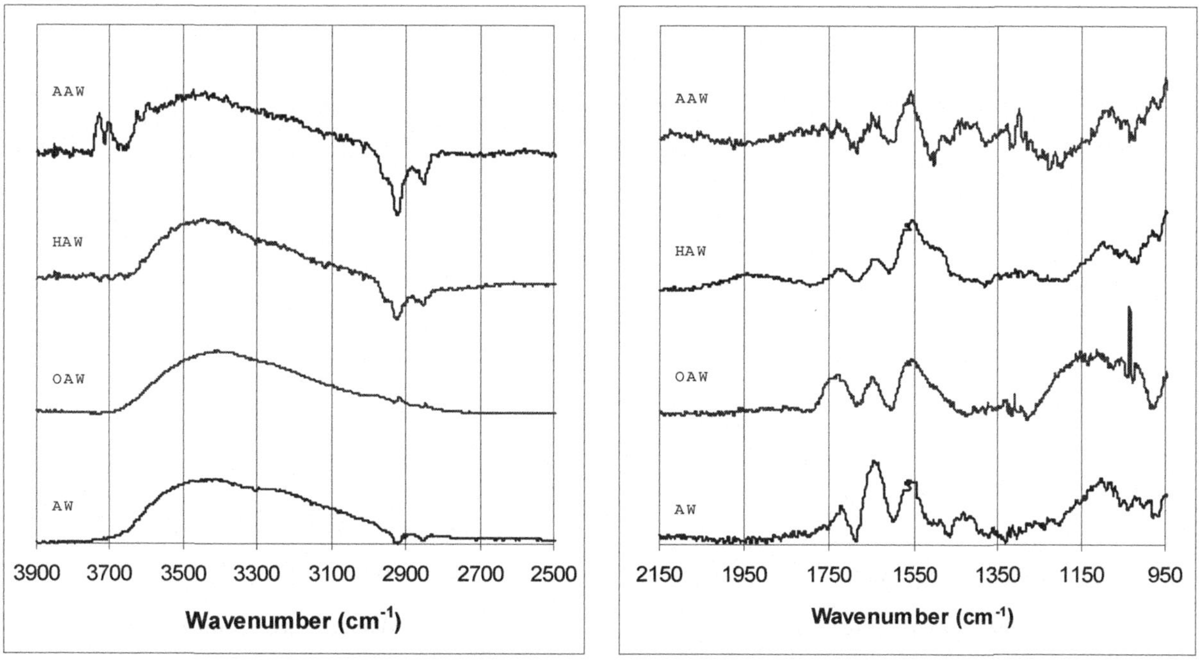

Figure 4.4 IR spectra for ACF10

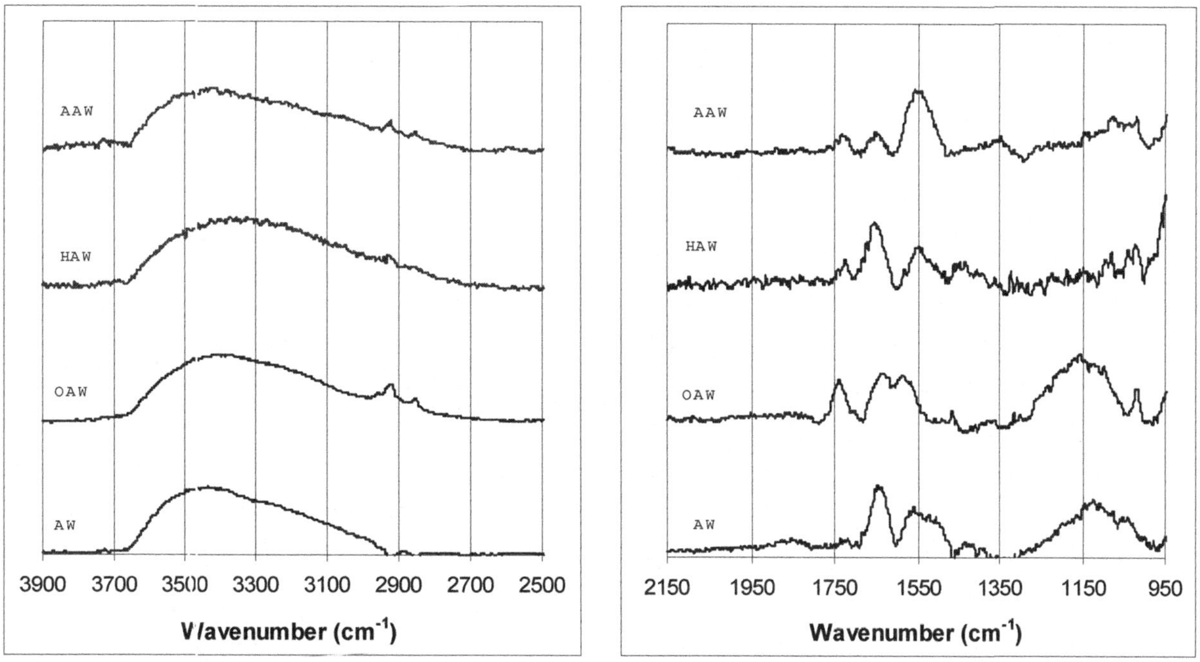

Figure 4.5 IR spectra for ACF15

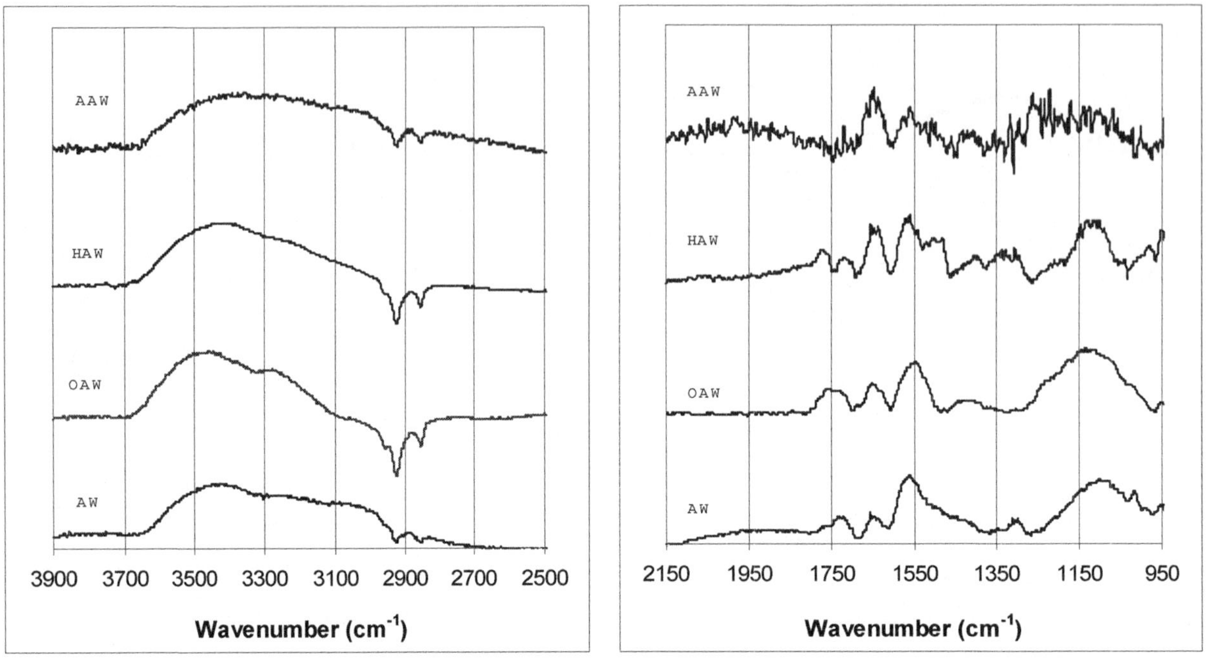

Figure 4.6 IR spectra for ACF20

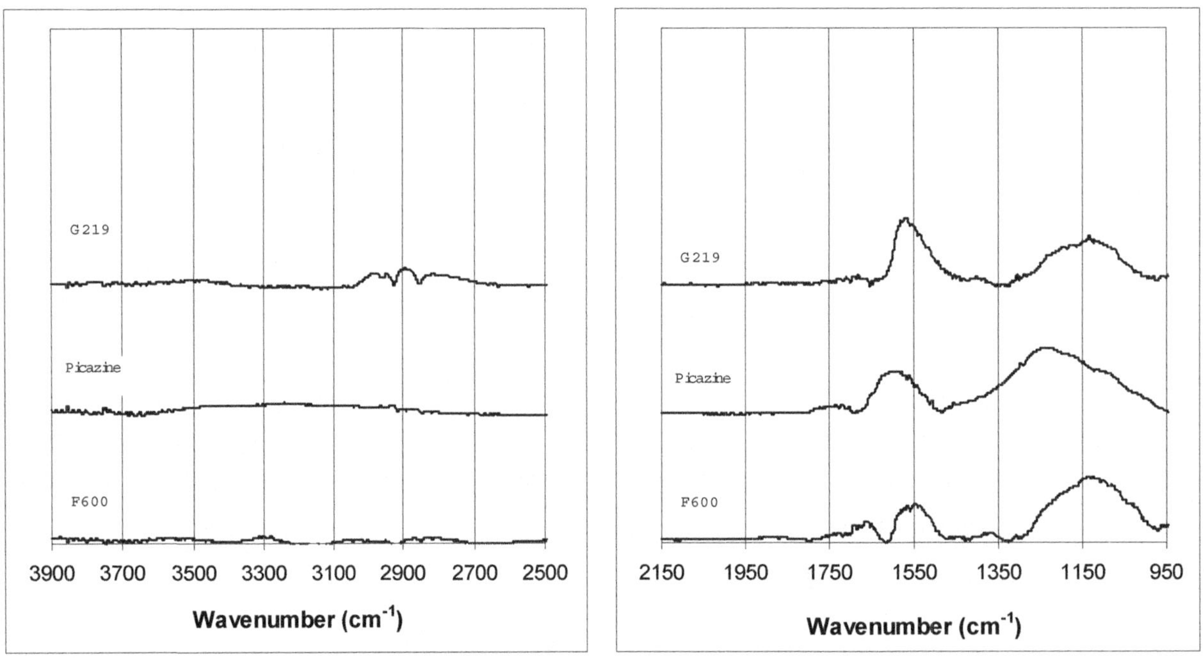

Figure 4.7 IR spectra for pulverized GACs

Table 4.4 Peak assignments for IR spectra.

Moiety	Peak	Reference
CO_2	3730, 3710 cm^{-1}	Lambert 1998
Bonded O-H	3100~3600 cm^{-1}	Lambert 1998
Germanium crystal Aliphatic CH_3, CH_2 stretching	2930, 2860 cm^{-1}	Spectra-Tech[a] Biniak et al. 1997 Lambert 1998
Carboxylic acids, ester, lactone	1720-1790 cm^{-1}	Meldrum and Rochester 1990 Nakahara and Sanada 1995 Sellitti et al. 1990
Quinones, ion radical structures C=O stretching	1650 cm^{-1}	Meldrum and Rochester 1990 Nakahara and Sanada 1995 Sellitti et al. 1990 Biniak et al. 1997
Ketones, esters, keto-enol C=O stretching	1570 cm^{-1}	Biniak et al. 1997
Deformation vibrations of surface –OH groups, in-plane C-H vibrations in C=C-H structure	1450-1300 cm^{-1}	Biniak et al. 1997
Carboxylic acids, esters, C-O-C stretching	1300-1000 cm^{-1}	Biniak et al. 1997 Lambert 1998

[a] Information provided with germanium crystal.

In the range of 3600-3000 cm^{-1}, a broad O-H stretching vibration was observed for all ACF samples (Figures 4.4-4.6). The width and asymmetry of this band indicates the presence of strong hydrogen bonds. These hydrogen bonds could be a result of interactions between N, H, and O containing moieties on the ACF surface or a result of strongly adsorbed water on hydrophilic functional groups on the ACF surfaces. In contrast, the broad O-H stretching vibration in the range of 3600-3000 cm^{-1} was not observed for GACs (Figure 4.7). Given that the ACFs were prepared from a phenolic resin, it is therefore likely that the broad O-H stretching vibration in the range of 3600-3000 cm^{-1} resulted from the presence of phenolic O-H groups. A comparison between ammonia-treated and other ACFs shows the same band shapes in the 3600-3000 cm^{-1} range, suggesting that N-H groups, which also absorb in this range (Lambert 1998), were not formed during ammonia-treatment or their presence was overshadowed by the presence of O-H stretching vibrations.

For the ACF samples, dominant peaks in the 2000-1500cm^{-1} range are observed at the 1720-1750, 1650, and 1550-1570 cm^{-1} bands, which can be attributed to C=O moieties in carboxylic acids, esters, and lactones (at 1720-1750 cm^{-1}); quinone structures (at 1650 cm^{-1}); and conjugated systems such as diketone, ketoester, and keto-enol structures (at 1570 cm^{-1}) (Meldrum and Rochester 1990, Sellitti et al. 1990, Nakahara and Sanada 1995). Peaks in this range may also be related vibrations of the condensed aromatic ring system. The 1750 cm^{-1} peak of HAW20 is split into two peaks (1778 and 1734 cm^{-1}), which may be attributable to the presence of acid anhydrides. IR spectra for ammonia-treated samples did not exhibit a peak around 1610 cm^{-1}, which would be characteristic for pyridine-type functional groups (Mangun et al. 2001). Although the pH_{PZC} data clearly indicated that ammonia-treatments led to the incorporation of basic surface functionalities, the employed IR technique was not able to prove the presence of nitrogen-containing functional groups in ammonia-treated ACFs.

IR spectra for pulverized F600 and G219 GACs exhibited peaks at the 1550-1570 cm^{-1} bands (Figure 4.7), which may again be attributable to the presence of diketone, ketoester, and keto-enol structures. In addition, F600 exhibited a peak at the 1650 cm^{-1} band, indicating the presence of quinone and /or ion-radical structures. The dominant peak in the 2000-1500cm^{-1} range for Picazine, a strongly acidic carbon, was observed at the 1600 cm^{-1} band, which may indicate the presence of COO^- salts (Starsinic et al. 1983). COO^- salts may have formed from COOH groups during the pulverization process, in which pulverized GAC was washed through a 45-μm sieve. Although deionized water was used for this purpose, metals may have eluted from the acidic Picazine carbon. Because the carbon was subsequently dried in the wash water, the formation of COO^- salts may have occurred during the drying process.

A number of ACF samples exhibited smaller bands in the 1450-1300 cm^{-1} range. These bands consist of a series of overlapping absorption bands that can be attributed to C-O stretching in O-C=O structures, the deformation vibration of surface hydyoxyl groups, in-plane C-H vibrations and C-N vibrations in heterocyclic structures. For both ACFs and GACs, the overlapping peaks that form a broad absorption band in the 1300-1000 cm^{-1} range can be assigned to ether, epi-oxide and phenolic structures (Figures 4.4-4.7).

Compared to all other ACFs, oxidized ACFs exhibited an increase in the intensities of bands at 1720-1750 cm^{-1} and 1300-1000 cm^{-1}. Stronger signals in these two regions suggest an increase in carboxylic acid functionalities, a result that is consistent with the larger oxygen content and the greater acidity of the oxidized ACFs (Table 4.2). The intensity of the band in the 1300-1000 cm^{-1} region decreased in both hydrogen-treated and ammonia-treated ACF samples. This result suggests that oxygen-containing functional groups were lost during hydrogen- and ammonia-treatments, which again is consistent with the results of the elemental analysis data and the loss of acidic functional groups. It needs to be pointed out, however, that IR spectra of ACFs with different activation levels that had undergone the same surface chemistry modification were not identical. Thus, subtle differences in surface chemistry may have existed in ACFs that had undergone the same surface chemistry modification even though their elemental composition and pH_{PZC} were similar.

SINGLE-SOLUTE TCE AND MTBE ISOTHERMS

To elucidate the effects of adsorbent surface chemistry and pore structure on the adsorption of organic pollutants from aqueous solution, single-solute TCE and MTBE isotherms experiments were conducted with each member of the ACF matrix. To compare single-solute isotherm results in a quantitative manner, isotherms were described by the Freundlich model [$q_e = K\,(C_e)^{1/n}$], where q_e and C_e are the equilibrium solid-phase and liquid-phase concentrations, respectively, and K and 1/n are fitting parameters. A larger K value represents a larger adsorption capacity while a larger 1/n value represents a more homogeneous adsorbent with a narrower site energy distribution (Derylo-Marczewska et al. 1984, Carter et al. 1995).

Figure 4.8 summarizes TCE and MTBE isotherm data along with Freundlich isotherm model fits for hydrogen-treated ACFs, and Table 4.5 summarizes TCE and MTBE Freundlich isotherm parameters for all adsorbents. As indicated in Figure 4.8a and by the K values in Table 4.5, TCE adsorbed to a greater extent than MTBE. Given the greater aqueous solubility of MTBE, primarily because its ether oxygen can serve as a hydrogen bond acceptor in water, MTBE adsorption on activated carbon surfaces requires the disruption of relatively strong solute/solvent interactions. Consequently, adsorption of MTBE is energetically less favored than adsorption of TCE, for which solute/solvent interactions are weaker. As illustrated in Figure 4.8b, differences in TCE and MTBE adsorption capacities were largely attributable to differences in the aqueous solubility of the adsorbates, and these results are consistent with well-established adsorbate solubility or polarity effects on adsorption (e.g., Lundelius 1920, Weber 1972, McGuire et al. 1978, Belfort 1979, Crittenden et al. 1999). Nonetheless, the difference between the TCE and MTBE isotherms on HAW10 (Figure 4.8b) illustrates that factors other than adsorbate solubility affected the adsorption of TCE and MTBE.

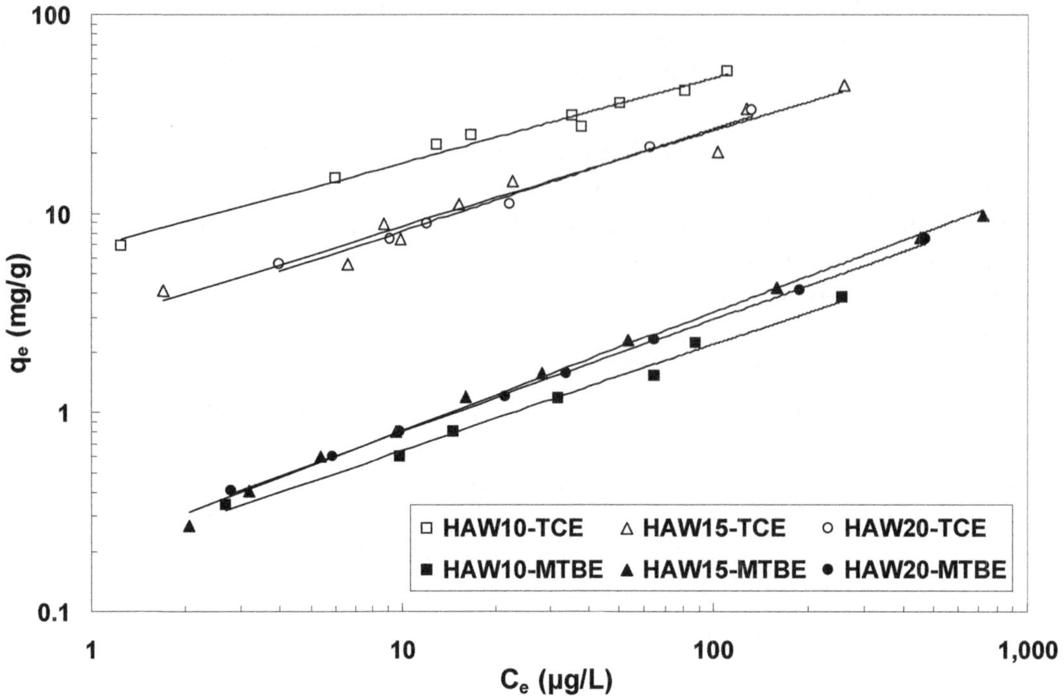

Figure 4.8a TCE and MTBE isotherms on hydrogen-treated ACFs. C_e in µg/L.

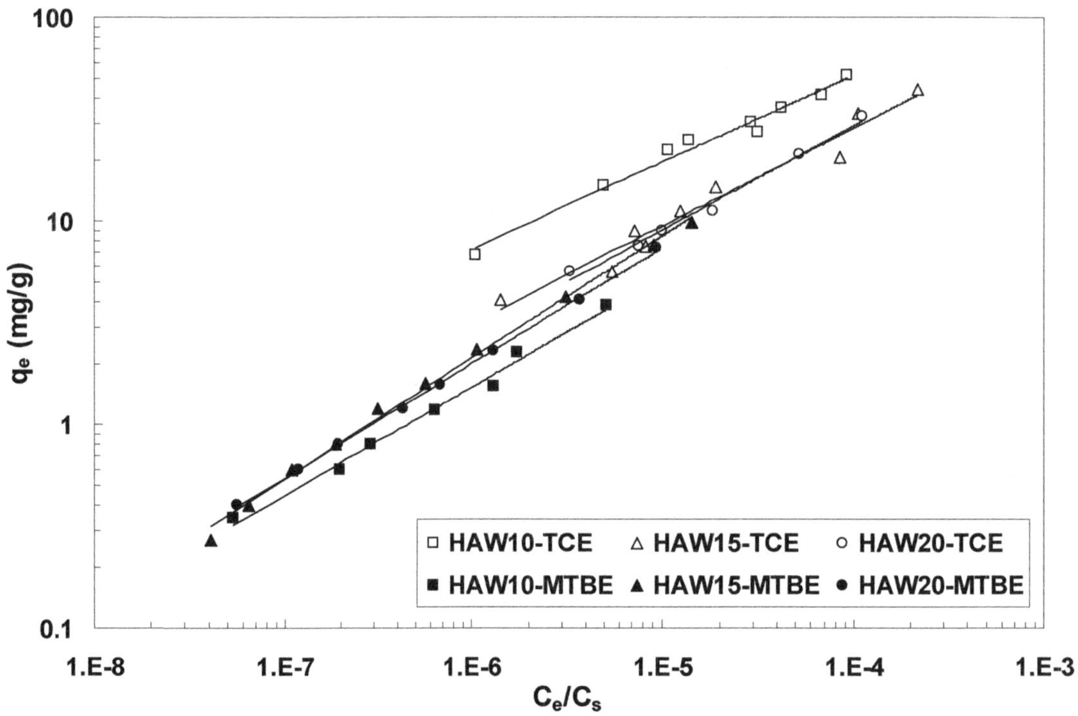

Figure 4.8b TCE and MTBE isotherms on hydrogen-treated ACFs. C_e normalized with respect to aqueous solubility of adsorbates (C_s).

Table 4.5 Freundlich isotherm parameters for single-solute TCE and MTBE isotherms. The 95% confidence intervals for each parameter are shown in parentheses.

Carbon	TCE		MTBE	
	K* Average (Lower; Upper)	1/n Average (Lower; Upper)	K* Average (Lower; Upper)	1/n Average (Lower; Upper)
AW10	5.23 (3.95; 6.92)	0.487 (0.403; 0.570)	0.0886 (0.0785; 0.1000)	0.603 (0.572; 0.633)
AW15	2.11 (1.82; 2.43)	0.554 (0.511; 0.596)	0.141 (0.128; 0.155)	0.634 (0.608; 0.660)
AW20	1.56 (1.35; 1.81)	0.593 (0.545; 0.641)	0.103 (0.0947; 0.112)	0.672 (0.646; 0.697)
OAW10	1.19 (1.04; 1.35)	0.604 (0.572; 0.637)	0.0451 (0.0413; 0.0493)	0.589 (0.571; 0.606)
OAW15	0.627 (0.494; 0.798)	0.591 (0.534; 0.649)	0.0376 (0.0342; 0.0412)	0.630 (0.611; 0.648)
OAW20	0.677 (0.584; 0.786)	0.595 (0.559; 0.632)	0.0301 (0.0241; 0.0375)	0.664 (0.610; 0.718)
HAW10	6.75 (5.67; 8.05)	0.426 (0.374; 0.479)	0.190 (0.152; 0.238)	0.532 (0.471; 0.594)
HAW15	2.85 (2.10; 3.86)	0.482 (0.393; 0.570)	0.206 (0.183; 0.232)	0.597 (0.566; 0.627)
HAW20	2.51 (1.95; 3.23)	0.514 (0.436; 0.592)	0.220 (0.206; 0.234)	0.564 (0.547; 0.581)
AAW10	4.61 (3.77; 5.64)	0.485 (0.426; 0.544)	0.164 (0.145; 0.185)	0.562 (0.531; 0.592)
AAW15	1.98 (1.53; 2.55)	0.567 (0.483; 0.651)	0.197 (0.170; 0.229)	0.546 (0.510; 0.582)
AAW20	2.05 (1.77; 2.38)	0.512 (0.467; 0.556)	0.115 (0.0986; 0.134)	0.631 (0.589; 0.674)
F600	4.05 (3.50; 4.69)	0.462 (0.422; 0.502)	0.150 (0.142; 0.159)	0.486 (0.475; 0.498)
G219	2.63 (2.08; 3.32)	0.484 (0.426; 0.543)	0.127 (0.112; 0.143)	0.540 (0.515; 0.564)
Picazine	0.099 (0.081; 0.121)	0.616 (0.563; 0.668)	0.00193 (0.00150, 0.00248)	0.841 (0.797, 0.884)

* Units of K: $(mg/g)(L/\mu g)^{1/n}$

Pore Structure Effects

With respect to pore structure, the results in Figure 4.8 show that the ranking of the ACFs in terms of adsorption capacity differed for the two solutes. For TCE, HAW10 exhibited the largest adsorption capacity even though it had the smallest BET surface area of the hydrogen-treated ACFs. The same trend was observed for all other surface chemistry modifications as indicated by the Freundlich K-values in Table 4.5. In comparison, MTBE adsorption capacities were relatively similar at all three ACF activation levels (Figure 4.8). The TCE isotherm data therefore suggest that ACF10 had a larger pore volume in a size range suitable for TCE adsorption than ACFs 15 and 20. Similarly, the MTBE isotherm data suggest that ACFs at all three activation levels had similar pore volumes in a size range suitable for MTBE adsorption. Both TCE and MTBE isotherm data clearly illustrate that adsorption capacities cannot be predicted from the BET surface areas or total micropore volumes (Table 4.1).

To assess in which pore sizes TCE and MTBE adsorb preferentially, adsorption capacities at an equilibrium liquid-phase concentration of 50 μg/L ($q_{e,\ 50}$) were correlated with ACF pore volumes in a given size range for each surface chemistry modification. Following a trial-and-error approach similar to that described by Ebie et al. (1995), the results showed that TCE adsorption was controlled by pores in the 7-10 Å width range while MTBE adsorption was controlled by pores in the 8-11 Å width range. These results suggest that both TCE and MTBE adsorb primarily in pores with widths that are 1.3 to 1.8 times larger than their kinetic diameters. Considering that the second-largest dimension of both adsorbates is similar (Figure 3.1), the results of the current study do not fully substantiate that adsorption of organic compounds from aqueous solutions is controlled by the second-largest adsorbate dimension as suggested by Kasaoka et al. (1989b). Given that pores of the studied ACFs are elliptical (Daley et al. 1996), it appears reasonable that the flat TCE molecule is able to access pores with a smaller dimension along the minor axis than the MTBE molecule, which approximates a tetrahedron. Thus, pores accessible to TCE corresponded to pores with smaller widths, as calculated by the DFT, which assumes slit-shaped pores.

For TCE, Figure 4.9 illustrates that the available pore volume in the 7-10 Å diameter range primarily controlled TCE adsorption on the more hydrophobic ACFs (AW, HAW, AAW) even though the oxygen contents of these carbons varied by a factor of about three (~1.5 - 4.5%). A comparison among Freundlich K-values (Table 4.5), representing TCE adsorption capacities at an equilibrium liquid-phase concentration of 1 μg/L, substantiated that pore structure effects dominated over surface chemistry effects for the more hydrophobic ACFs; i.e., 95% confidence intervals generally overlapped for AW, HAW, and AAW at a given activation level while ACF 10 always exhibited the largest K-value for a given surface chemistry modification. Figure 4.9 also shows that the correlation established by the more hydrophobic ACFs predicted the adsorption capacity of the relatively hydrophobic F600 and G219 GACs reasonably well. For the oxidized ACFs, a separate correlation indicated that surface chemistry effects in addition to the availability of a suitable pore volume controlled TCE adsorption for ACFs with larger oxygen contents (~10%). The correlation established by the oxidized ACFs appropriately predicted the adsorption capacity of Picazine GAC with an oxygen content of about 16%.

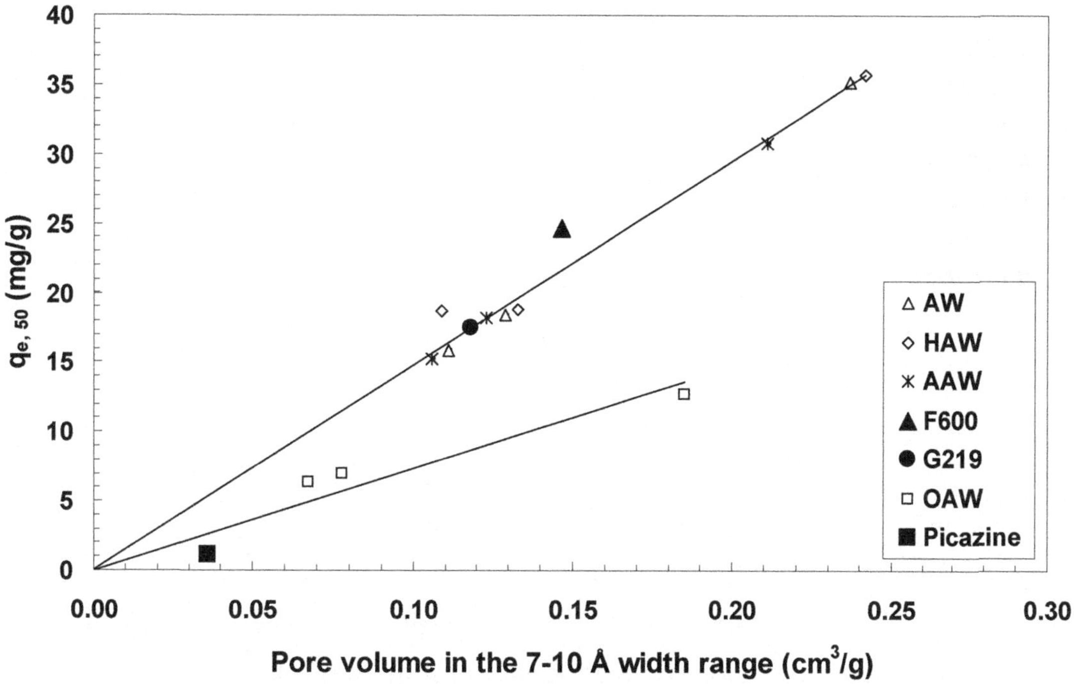

Figure 4.9 Effect of pore volume on TCE adsorption

A plot of MTBE adsorption capacity versus micropore volume in the 8-11 Å diameter range (Figure 4.10) showed a trend similar to that depicted in Figure 4.9 for TCE. However, surface chemistry effects were of some importance even among the more hydrophobic ACFs. A comparison of the Freundlich K-values in Table 4.5 also shows that the hydrogen-treated ACFs; i.e., the most hydrophobic ACFs of the matrix, exhibited the largest MTBE adsorption capacities at a given activation level. Nonetheless, differences among the more hydrophobic ACFs at a given activation level were small compared to the decrease in MTBE adsorption capacity that was observed upon oxidation (Table 4.5).

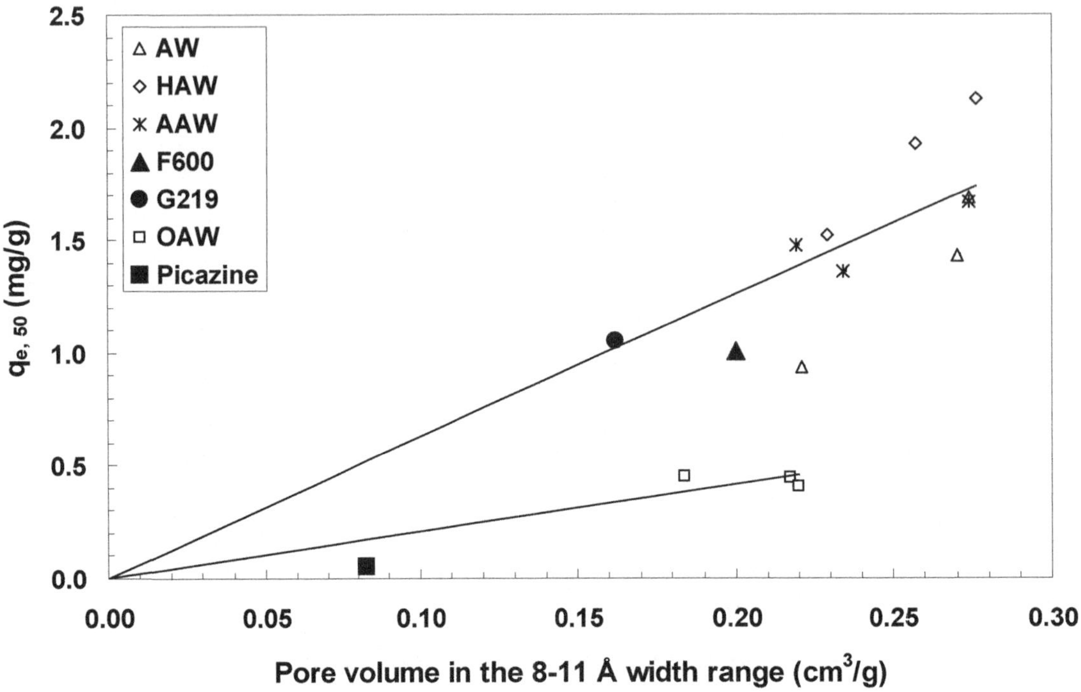

Figure 4.10 Effect of pore volume on MTBE adsorption

Surface Chemistry Effects

Overall, the isotherm data showed that the adsorption of both relatively hydrophobic (TCE) and relatively hydrophilic (MTBE) organic pollutants from aqueous solution is adversely affected by the introduction of surface oxygen groups (see Freundlich K-values in Table 4.5). Prior studies have shown that the adsorption of polar organics such as acetone (Mangun et al. 1999, Dimotakis et al. 1995) or acetaldehyde (Mangun et al. 1999) from the gas-phase is enhanced upon surface oxidation because of dipole-dipole interactions (Mangun et al. 1999) or hydrogen bonding, and a similar result may be anticipated for MTBE vapor. However, in aqueous solution, MTBE has to compete with water for polar adsorption sites. Given that the concentration of MTBE is small compared to that of water, water adsorption should be favored over MTBE adsorption at polar surface sites.

To verify the importance of solvent/adsorbent interactions, MTBE and TCE adsorption was also studied from cyclohexane, a nonpolar solvent. In contrast to the results obtained in aqueous solution, Figure 4.11 shows that the oxidized ACF exhibited the largest MTBE adsorption capacity (by a factor of about 5 to 6) when the solvent was cyclohexane, and the observation was reproducible for the two tested adsorbent doses.

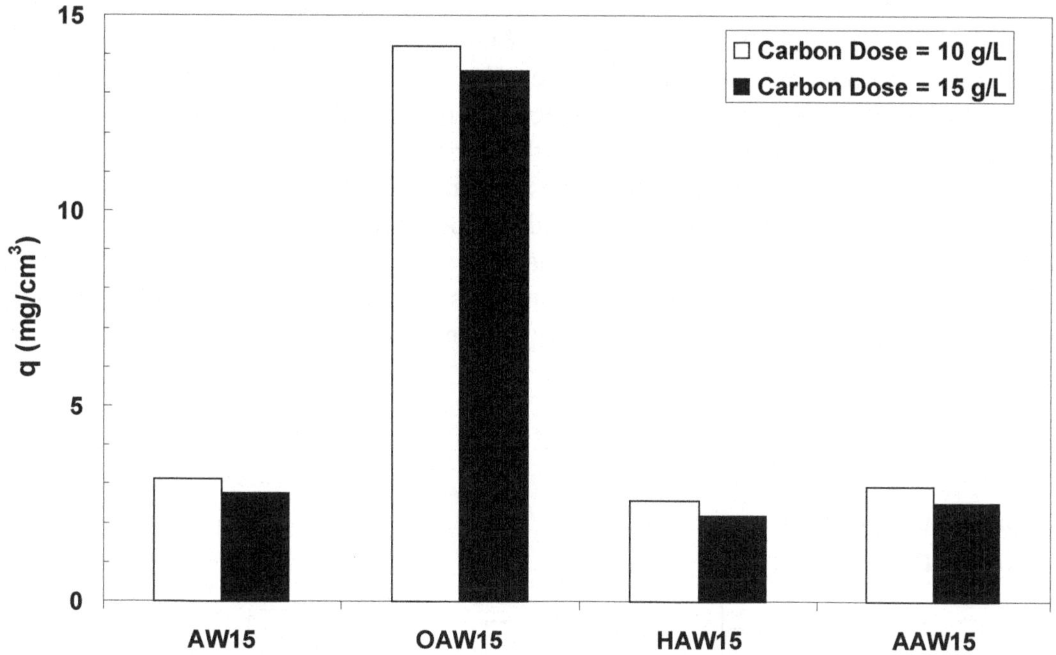

Figure 4.11 Effect of chemical treatments on MTBE adsorption from cyclohexane. Solid-phase MTBE concentrations were normalized by the pore volume in the 8-11 Å width range.

The improved performance of the oxidized ACF in cyclohexane can be explained by the preferential adsorption of MTBE on oxygen-containing functional groups such as carboxylic acid and phenolic hydroxyl groups. In that case, hydrogen bonds can form between the hydrogen atoms of these surface groups and the ether-oxygen of MTBE. However, in the presence of water, hydrogen bonds will preferentially form between water and these surface groups. Consequently, the availability of specific adsorption sites; i.e., hydrogen-bond donor sites, for MTBE will decrease greatly in aqueous solution. However, the loss of such sites to water adsorption alone cannot explain why the MTBE adsorption capacity of oxidized ACFs in aqueous solution is lower than that of more hydrophobic ACFs with a smaller surface concentration of hydrogen-bond donor sites. It is plausible that hydrogen-bond donor sites become essentially unavailable to MTBE in aqueous solution and that MTBE adsorption therefore shifts from a site-specific hydrogen-bonding mechanism in cyclohexane to a non-specific dispersive mechanism in aqueous solution, where MTBE adsorbs instead on the graphitic basal planes of the adsorbent. Hence, the lower MTBE adsorption capacity of oxidized adsorbents with a larger concentration of surface-oxygen groups results from the increased formation of water clusters that reduce access to the graphitic basal planes, reduce the interaction energy between MTBE and the adsorbent surface, and/or block pore entrances (Coughlin and Ezra 1968, Mahajan et al. 1980, Kaneko et al. 1989, Pendleton et al. 1997, Franz et al. 2000, Müller et al. 2000).

Results for TCE adsorption from cyclohexane showed that surface chemistry effects were negligible (Figure 4.12), a result that suggests that TCE adsorption occurs by a non-specific dispersive mechanism. In aqueous solution, however, the TCE adsorption capacity of oxidized adsorbents was compromised because of the enhanced formation of water clusters as explained above for MTBE. The results and interpretations of the current study are consistent with those of a recent study by Franz et al. (2000) who investigated the adsorption of phenol, aniline, nitrobenzene, and benzoic acid from both water and cyclohexane. Overall, the above results as well as those by Franz et al. (2000) show that enhanced water adsorption on polar adsorbents negates the benefits of providing specific adsorption sites capable of forming hydrogen-bonds with a targeted adsorbate such as MTBE (this study), phenol, aniline, or benzoic acid (Franz et al. 2000).

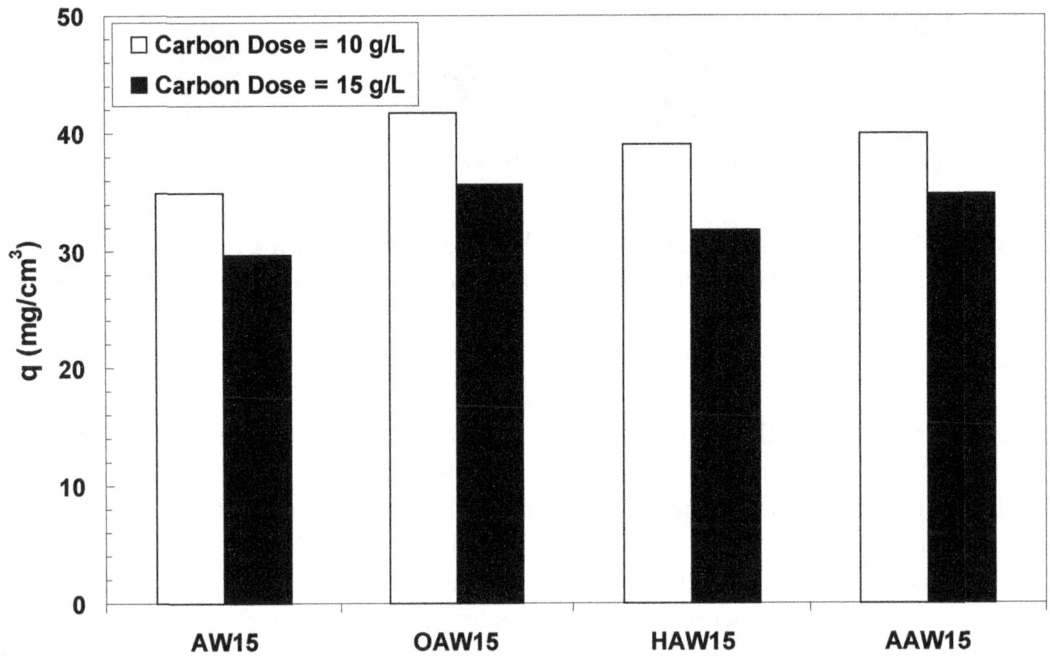

Figure 4.12 Effect of chemical treatments on TCE adsorption from cyclohexane. Solid-phase TCE concentrations were normalized by the pore volume in the 7-10 Å width range.

To determine the effects of adsorbent polarity on TCE and MTBE adsorption, adsorption capacities at an equilibrium liquid-phase concentration of 50 µg/L ($q_{e,\,50}$ in mg/g) were first normalized by the pore volume in the pore width range that controlled the adsorption of the targeted micropollutant (i.e., 7-10 Å for TCE, 8-11 Å for MTBE). The thus normalized micropollutant adsorption capacities ($q'_{e,\,50}$ in mg micropollutant/cm^3 pore volume) were then plotted against the sum of the oxygen and nitrogen contents (mmol/g) of each adsorbent as shown in Figures 4.13 and 4.14 for TCE and MTBE, respectively. Figure 4.13 illustrates that the TCE adsorption capacity decreased with increasing adsorbent polarity, and a similar trend was observed for MTBE (Figure 4.14).

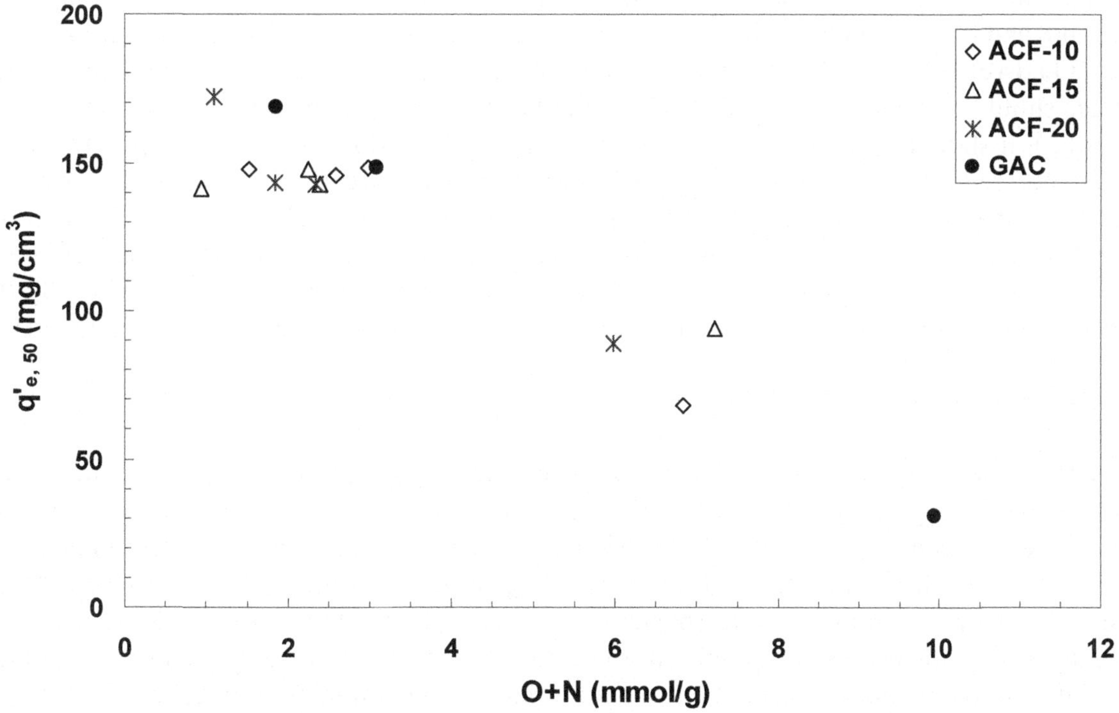

Figure 4.13 Effect of adsorbent hydrophobicity on TCE adsorption capacity

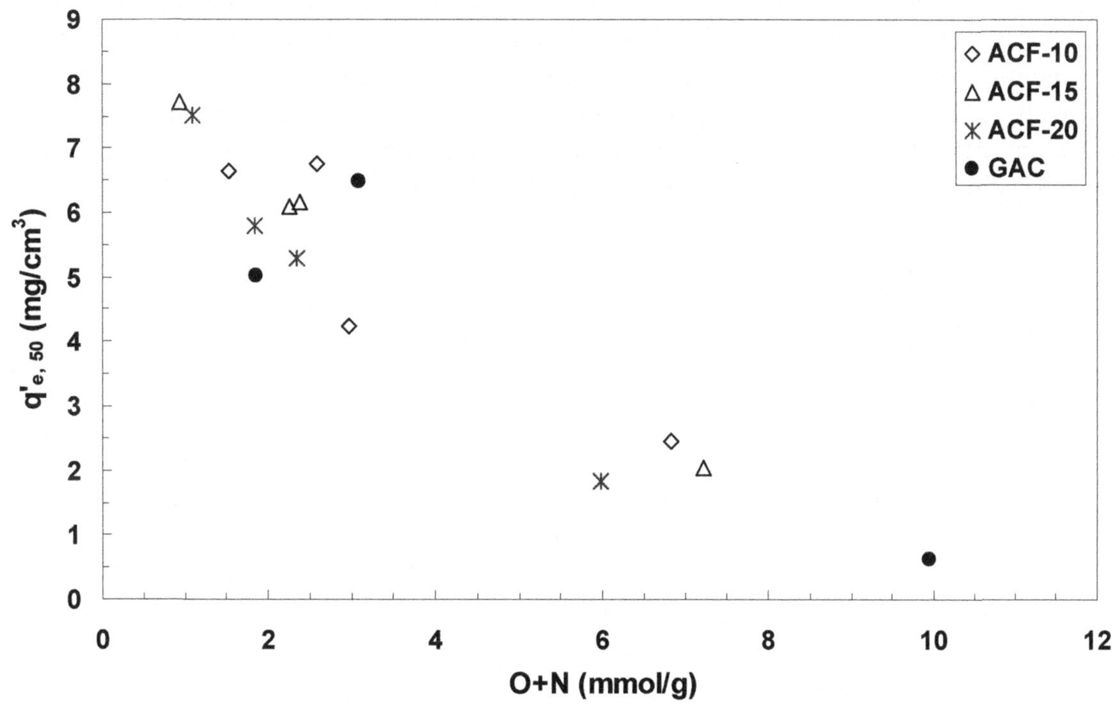

Figure 4.14 Effect of adsorbent hydrophobicity on MTBE adsorption capacity

The results in Figures 4.13 and 4.14 clearly illustrate that the effective removal of both hydrophobic and hydrophilic adsorbates from aqueous solution requires adsorbents that meet two criteria: (1) adsorbents should contain a large pore volume in a suitable pore size range, and (2) adsorbents should be hydrophobic; i.e., they should have a low oxygen and nitrogen content (< 2 to 3 mmol/g). It should be noted that surface chemistry modification techniques employed in this study did not reduce the polarity of the adsorbents to the extent that adsorbents were not effectively wetted by water. At this point, it is therefore not possible to define a minimum polarity, based on adsorbent oxygen and nitrogen contents, that assures adsorbents are effective in aqueous solutions. Figures 4.13 and 4.14 further illustrate that the trends established by the ACF matrix successfully described the adsorption behavior of the commercially available GACs.

NOM ADSORPTION ISOTHERMS

Figure 4.15 depicts DOC isotherms for acid-washed ACF samples. As shown in Figure 4.15, AW10 adsorbed negligible quantities of DOC. Given the narrow pores of AW10 (primarily 7-11 Å width), size exclusion effects can explain the lack of DOC adsorption on this ACF. In contrast, AW15 and AW20 exhibited larger DOC adsorption capacities, a result that is consistent with the presence of wider pores in these ACFs (primarily 9-13 Å width for both ACFs). Figure 4.15 also shows that DOC isotherms, when plotted in the form of $\log(q_e)$ versus $\log(C_e)$, do not yield the linear relationships generally observed for trace organic contaminants. Given that NOM is a mixture of polymeric compounds with a range of molecular weights, an isotherm model appropriate for heterodisperse polymer mixtures needs to be applied. As shown by Summers and Roberts (1988a), equilibrium liquid-phase concentrations (C_e in mg/L as C) should be normalized by the corresponding the activated carbon dose (D in g/L) before applying the Freundlich model. Figure 4.16 depicts dose-normalized DOC isotherms for the acid-washed ACF samples as well as the best fits of the corresponding Freundlich model [$q_e = K\,(C_e/D)^{1/n}$] to the experimental data. Figure 4.16 shows that the dose-normalized Freundlich model described the data well, illustrating that adsorption models derived for heterodisperse polymers are appropriate for describing DOC adsorption on activated carbons. Consequently, the resulting Freundlich K-values can be used to compare DOC adsorption capacities of the ACFs.

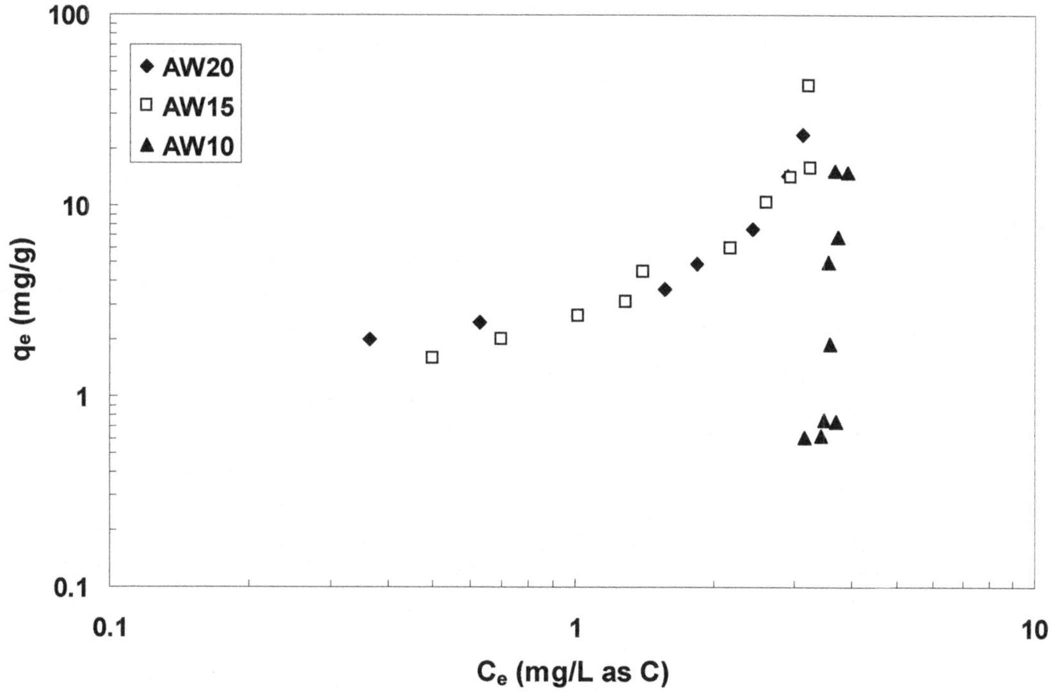

Figure 4.15 DOC isotherms for acid-washed ACFs

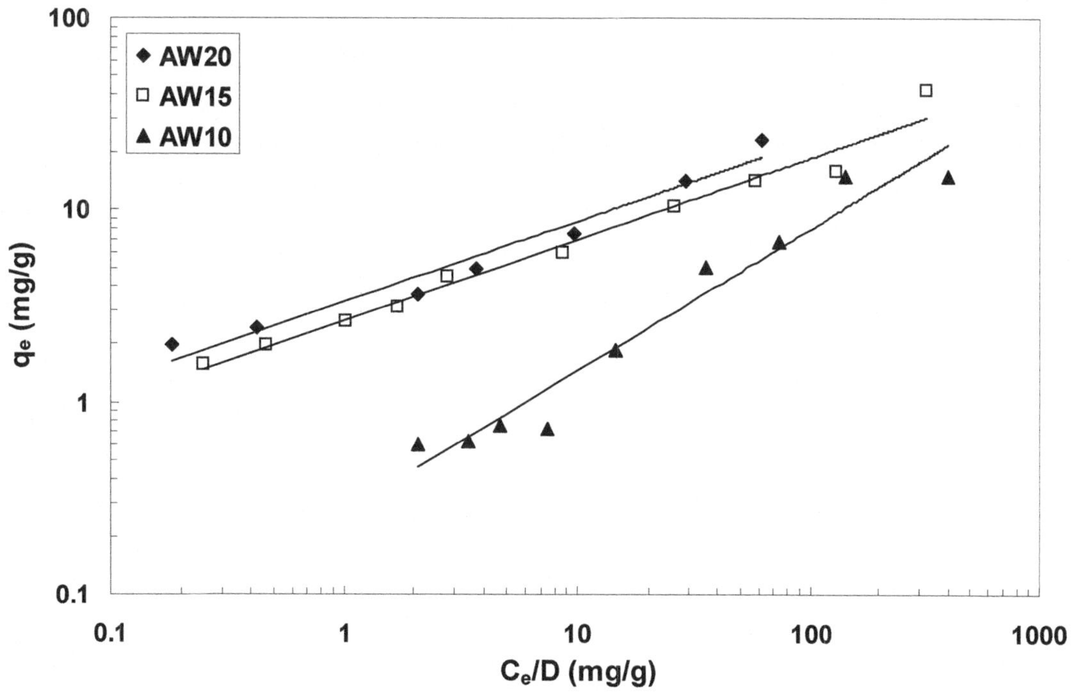

Figure 4.16 Dose-normalized DOC isotherms for acid-washed ACFs

As summarized in Table 4.6, AW10 had a significantly lower DOC adsorption capacity than AW15 and AW20. Similar obsevations were made when quantifying NOM adsorption by UV absorbance at 280 nm (Figures 4.17 and 4.18, Table 4.6).

Table 4.6 Freundlich isotherm parameters for dose-normalized DOC and UV_{280} isotherms for ACFs and pulverized GACs.

Carbon	Dose-Normalized DOC Isotherms		Dose-Normalized UV_{280} Isotherms	
	K $(mg/g)^{1-1/n}$	1/n	K $(L\ g^{-1}\ cm^{-1})^{1-1/n}$	1/n
AW10	0.269	0.735	.201	.897
AW15	2.65	0.423	.452	.492
AW20	3.33	0.419	.544	.540
OAW10	.469	.678	.402	.674
OAW15	1.21	.710	.750	.666
OAW20	1.09	.708	.578	.567
HAW10	.370	.570	.196	.782
HAW15	2.89	.477	.595	.479
HAW20	3.52	.406	.476	.489
AAW10	.268	.695	.149	.762
AAW15	3.58	.442	.500	.469
AAW20	2.69	.429	.566	.511
F600	3.60	.523	.842	.470
G219	7.16	.415	.916	.439
Picazine	2.74	.609	.981	.636

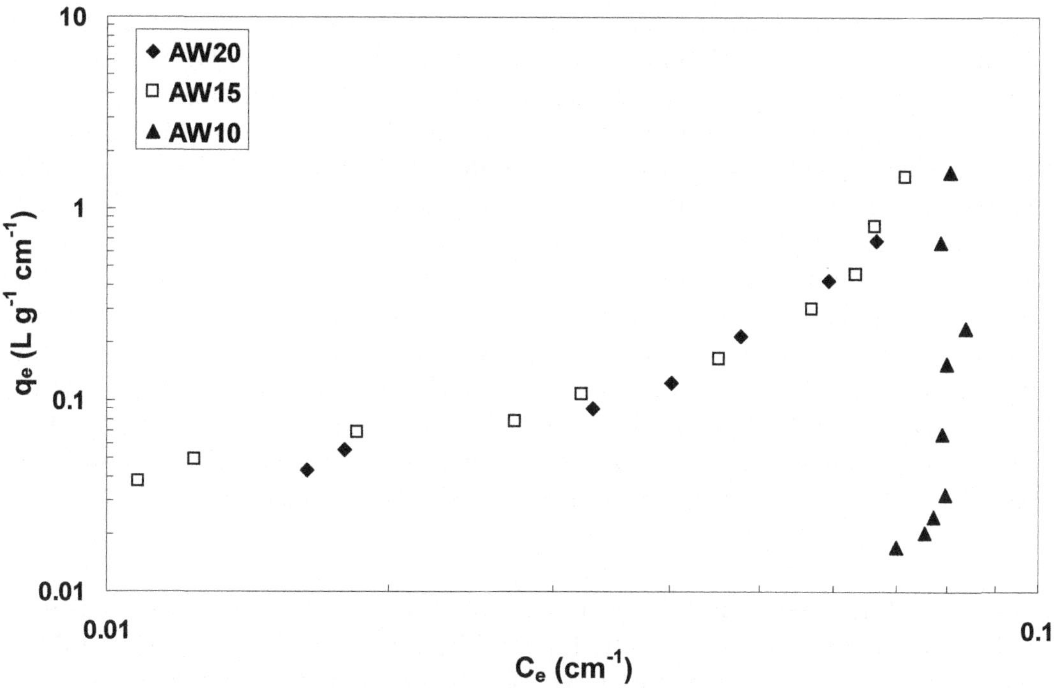

Figure 4.17 Isotherms of UV_{280}-absorbing NOM components for acid-washed ACFs

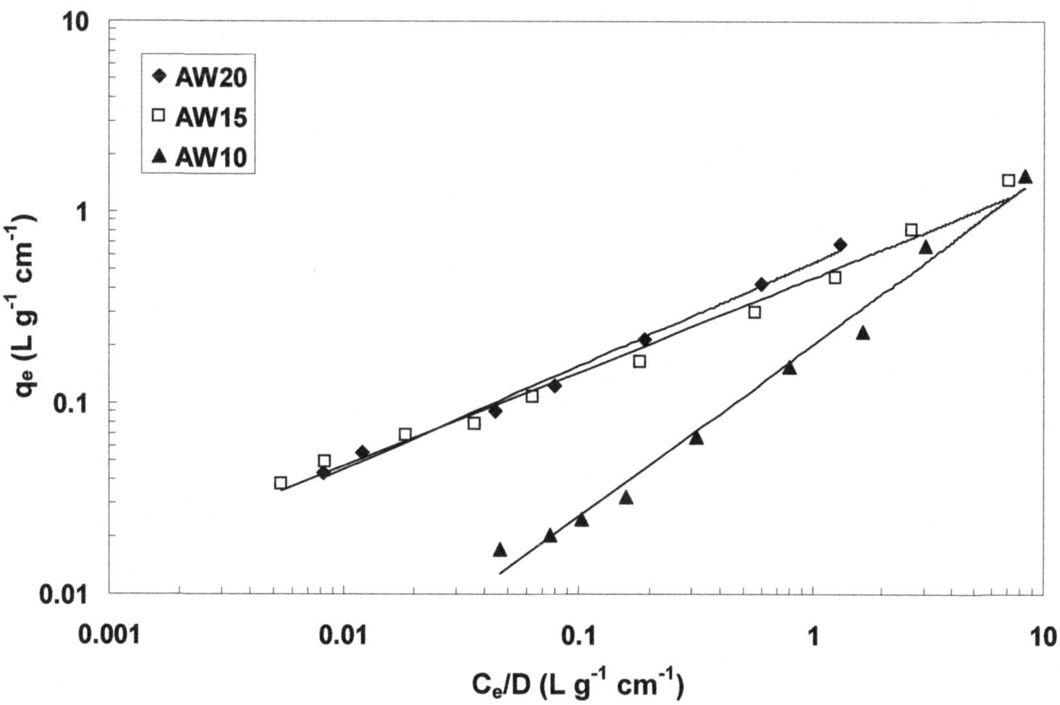

Figure 4.18 Dose-normalized isotherms of UV_{280}-absorbing NOM components for acid-washed ACFs

Figure 4.19 depicts dose-normalized DOC isotherms for the ACF10 members of the adsorbent matrix. The results in Figure 4.19 suggest that surface chemistry modifications had only a small effect on DOC adsorption isotherms. Slightly enhanced DOC adsorption on OAW10 may have been a result of a small increase in mesopore volume that resulted from the oxidation process. Figure 4.20 shows dose-normalized DOC isotherms for the ACF20 members of the adsorbent matrix. NOM adsorption capacities among the hydrophobic ACFs (i.e., AW, HAW, and AAW) were similar while a loss in adsorption capacity was observed for OAW20 at low C_e/D values or low surface loading levels. This result suggests that the negative surface charge of OAW20 adversely affected the adsorption of negatively charged NOM. Furthermore, enhanced water adsorption on the more polar oxidized ACF may have reduced the pore volume available to NOM as a result of pore blockage or pore constriction. However, at higher C_e/D values or higher surface loading levels, the NOM adsorption capacity of OAW20 was similar to that of the other ACFs. It is likely that adsorbed NOM controlled the ACF surface charge at higher loading levels, which explains why adsorption capacities were independent of surface chemistry modifications at higher C_e/D values. Results for ACF15 were similar to those depicted for ACF20 in Figure 4.20.

When quantifying NOM adsorption by UV absorbance at 280 nm, which targets the adsorption of aromatic and unsaturated NOM components, Figure 4.21 illustrates that OAW10 exhibited a larger adsorption capacity than the remaining ACF10 members of the adsorbent matrix. As explained above, a small increase in mesopore volume that resulted from the oxidation step may explain this result. For the ACF20 members of the adsorbent matrix, Figure 4.22 illustrates that the adsorption of UV_{280}-absorbing NOM components was not affected by ACF surface chemistry modifications.

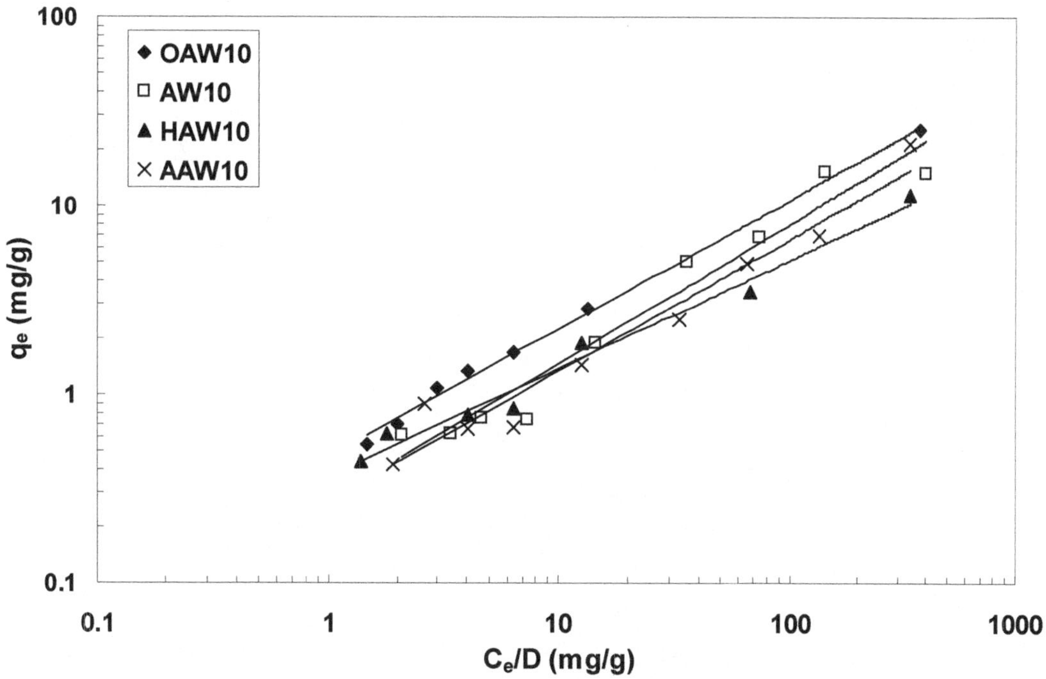

Figure 4.19 Effect of surface chemistry modifications on dose-normalized DOC isotherms for ACF10

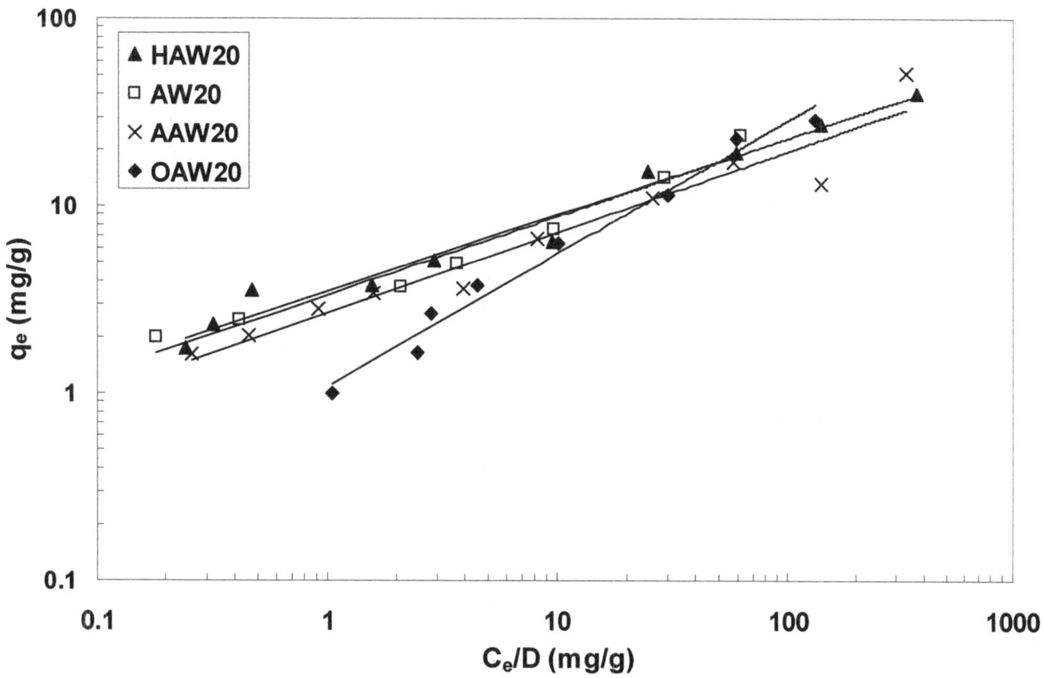

Figure 4.20 Effect of surface chemistry modifications on dose-normalized DOC isotherms for ACF20

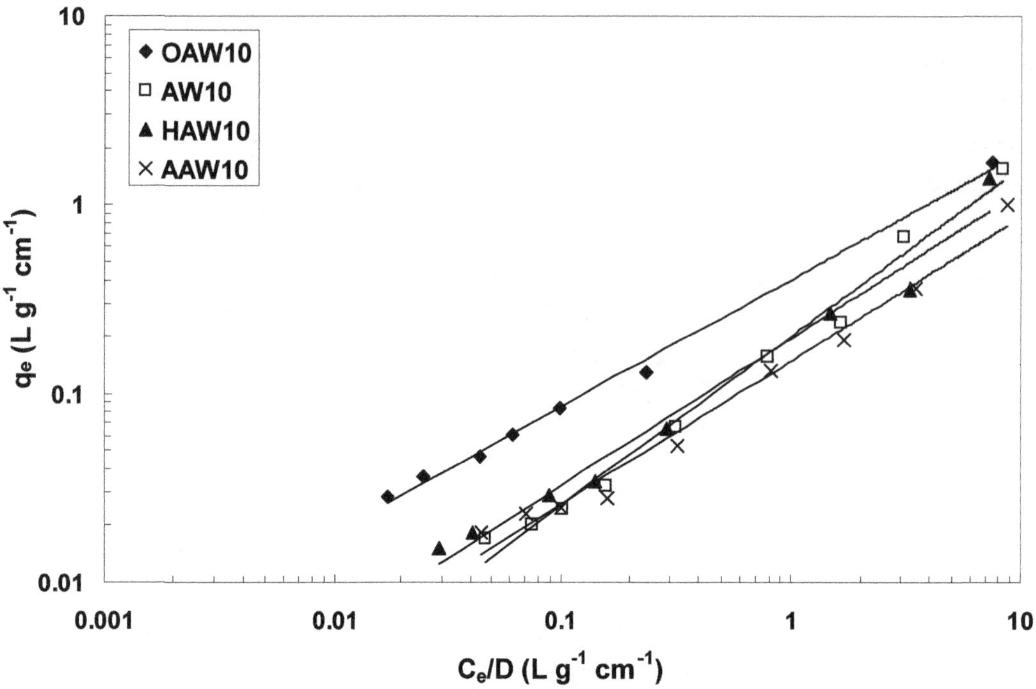

Figure 4.21 Effect of surface chemistry modifications on dose-normalized isotherms of UV_{280}-absorbing NOM components for ACF10

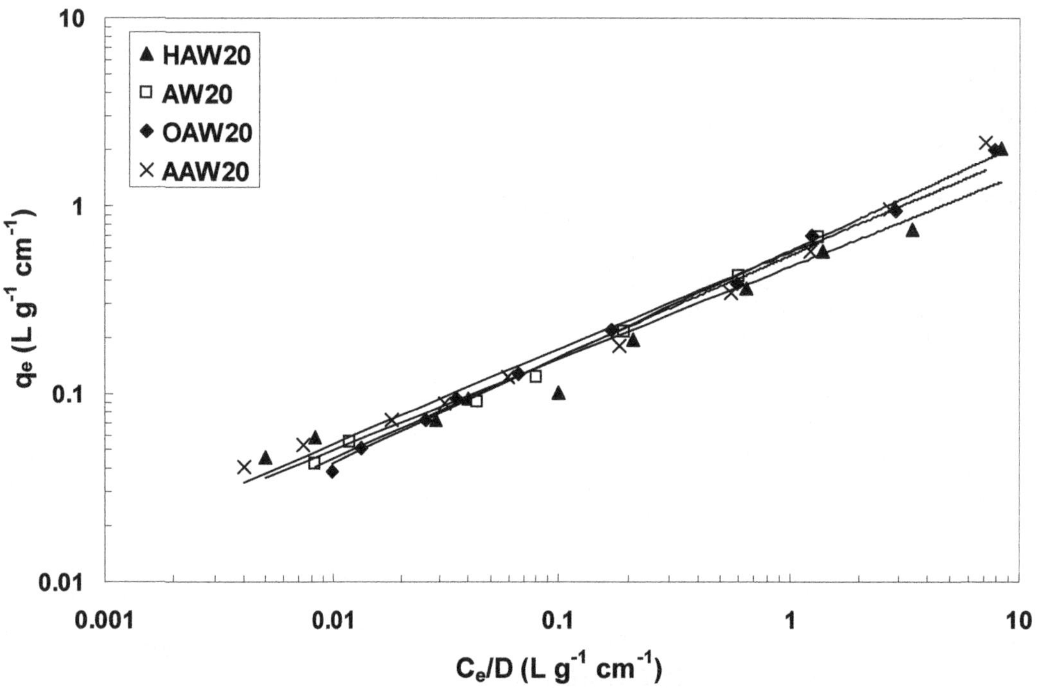

Figure 4.22 Effect of surface chemistry modifications on dose-normalized isotherms of UV_{280}-absorbing NOM components for ACF20

The contrasting results between Figures 4.20 (DOC isotherms) and Figure 4.22 (UV_{280} isotherms) suggests that the negative surface charge on OAW20 only reduced the adsorption of aliphatic NOM components; i.e., NOM components that do not adsorb UV light.

Figure 4.23 depicts dose-normalized DOC isotherms for the three commercially available GACs. Furthermore, Table 4.6 shows the Freundlich K and 1/n values that resulted from the regression analyses. The results in Figure 4.23 and the Freundlich parameters in Table 4.6 indicate that G219 exhibited the largest NOM adsorption capacity while that of Picazine and F600 was similar. The latter result was somewhat surprising given that the mesopore volume of Picazine GAC was approximately 3 times larger than that of F600 while the micropore volumes were similar for both GACs. Therefore, both adsorbent surface chemistry and pore structure affected the adsorption of DOC by the GACs.

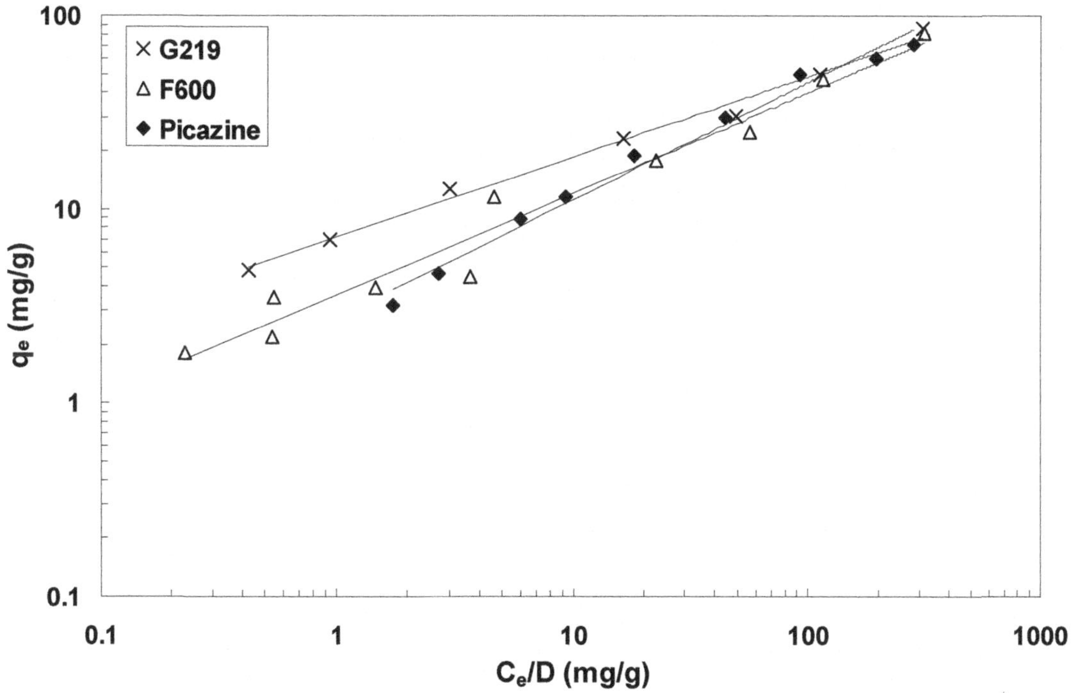

Figure 4.23 Dose-normalized DOC isotherms on pulverized GACs

To separate surface chemistry and pore structure effects, DOC and UV_{280} adsorption capacities of the hydrophobic ACFs (i.e., AW, HAW, and AAW ACFs) were correlated with pore volumes in a given size range. Following a trial-and-error approach similar to that described by Ebie et al. (1995), the results showed that DOC and UV_{280} adsorption were controlled by pores in the 12-500 Å width range (Figures 4.24 and 4.25). The observation that DOC adsorption occurs in mesopores and large micropores is consistent with the results of prior studies that investigated the adsorption of aquatic NOM (e.g., Ebie et al. 1995, Newcombe et al. 1997). Figure 4.24 illustrates that oxidized ACFs exhibited slightly smaller DOC adsorption capacities than the remaining ACFs. However, the same was not true when comparing adsorption capacities of UV_{280}-absorbing NOM components (Figure 4.25).

When plotting NOM adsorption capacities of the pulverized GACs, both Figures 4.24 and 4.25 illustrate that the strongly acidic Picazine GAC exhibited an NOM adsorption capacity that falls slightly below the line that best described NOM adsorption on the hydrophobic ACFs. This result may be attributable to greater electrostatic repulsion between negatively charged NOM molecules and the negatively charged GAC surface. In contrast, the basic F600 and G219 GACs exhibited NOM adsorption capacities that greatly exceeded those of the ACFs at a given pore volume. Given that the pH_{PZC} of ammonia-treated ACFs fell between those of G219 and F600, this result was somewhat unexpected. Further studies would be needed to measure the surface charge of each adsorbent at the pH of San Joaquin Delta water (SJDW). If both G219 and F600 exhibited a larger positive surface charge at the pH of SJDW than the ammonia-treated ACFs, the enhanced NOM adsorption capacity of G219 and F600 could be explained based on enhanced electrostatic attraction between the negatively charged NOM and the positively charged GAC surfaces.

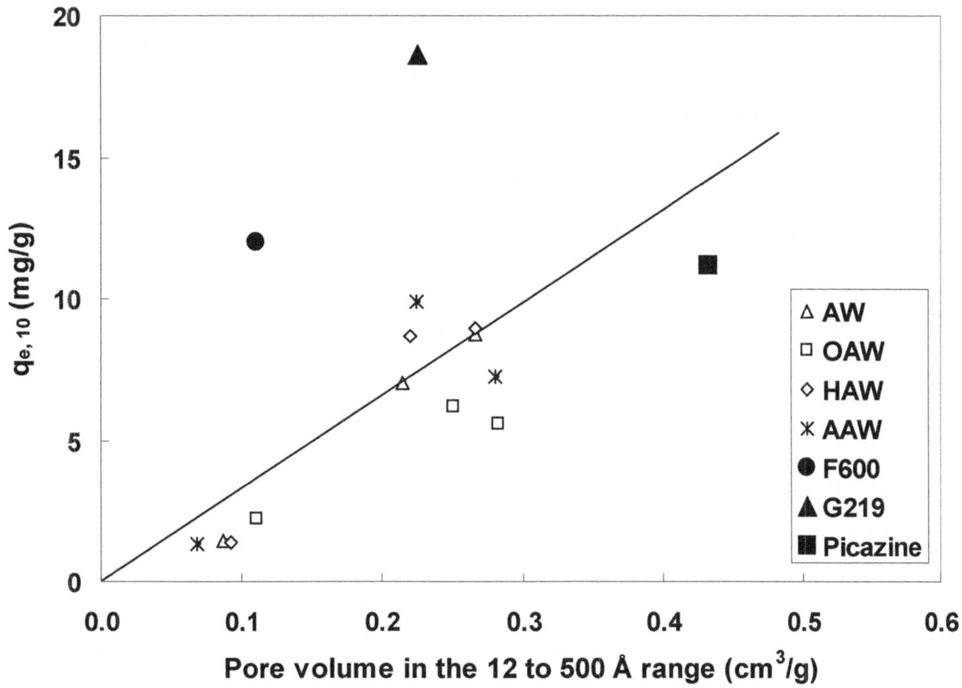

Figure 4.24 Correlation between the DOC adsorption capacity at C_e/D = 10 mg/g and pore volume in the 12-500 Å width range

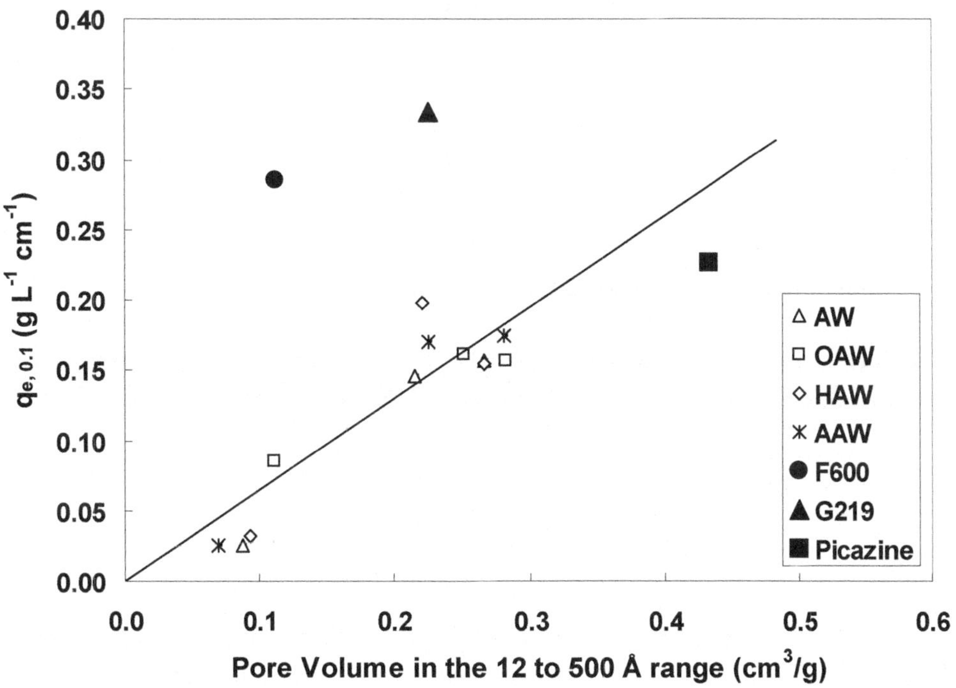

Figure 4.25 Correlation between the adsorption capacity of UV_{280}-absorbing NOM components at C_e/D = 0.1 g L^{-1} cm^{-1} and pore volume in the 12-500 Å width range

MICROPOLLUTANT ADSORPTION IN THE PRESENCE OF NOM

To evaluate the effects of competing NOM components on micropollutant adsorption capacities, MTBE and TCE isotherm experiments were performed in Sacramento-San Joaquin Delta water (SJDW). Figure 4.26 compares the TCE isotherm for AW20 in ultrapure water to that in SJDW. Figure 4.26 shows that competition between NOM and TCE lowered the TCE adsorption capacity. At an equilibrium liquid-phase concentration of 10 μg/L, the equilibrium solid-phase concentration of TCE decreased from 6.1 mg/g in ultrapure water to 2.6 mg/g in the presence of NOM, a capacity decrease of about 58%. In the presence of NOM, TCE and MTBE adsorption capacities of the adsorbents tested in this study decreased by 15-70%.

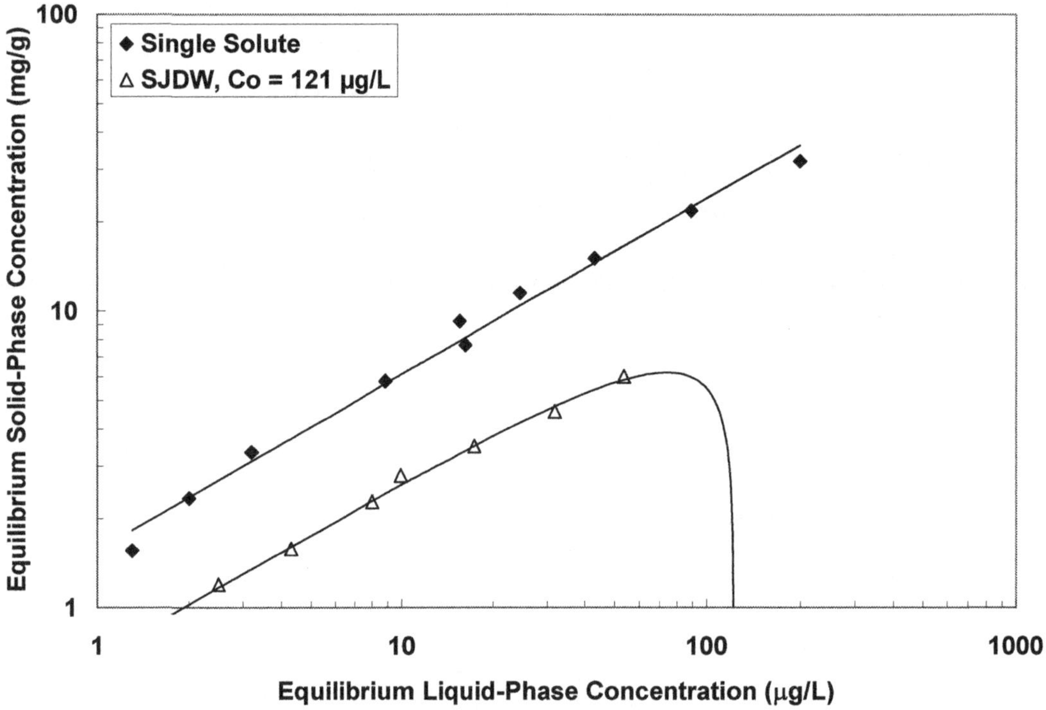

Figure 4.26 Effect of NOM on TCE adsorption by AW20. Solid line through single-solute data represents Freundlich isotherm fit. Solid line through SJDW data represents IAST model fit.

Pore Structure Effects

To assess the effects of adsorbent pore structure on TCE adsorption from SJDW, Figure 4.27 compares TCE isotherms for the three acid-washed carbons in an alternative format, where the percentage of TCE remaining in solution at equilibrium is plotted as a function of the applied ACF dosage. In the presence of competing NOM, micropollutant isotherms plotted in this format are not dependent on the initial micropollutant concentration, provided that the micropollutant is present at trace levels. The points in Figure 4.27 are the experimental data, and the lines through each data set represent the best fits of the IAST model.

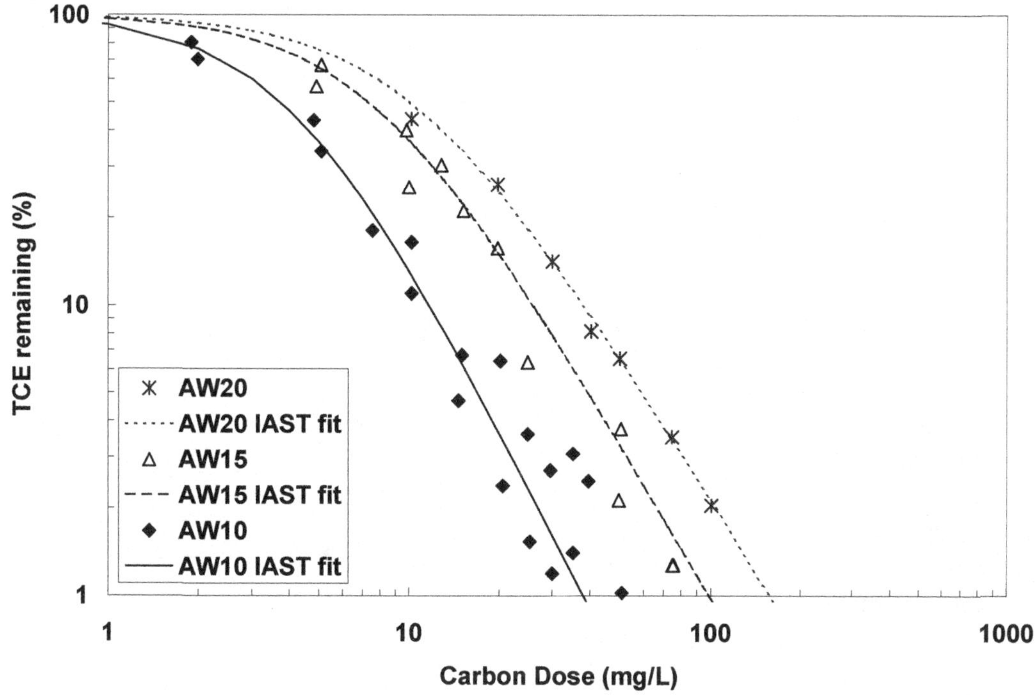

Figure 4.27 Effect of ACF pore structure on TCE adsorption from SJDW (acid-washed ACFs)

According to Figure 4.27, AW10 exhibited the largest adsorption capacity for TCE in SJDW while AW20 exhibited the lowest. This trend is the same as that observed for single-solute TCE isotherm experiments (Table 4.5). Given that AW10 had the lowest BET surface area, this result again shows that the BET surface area is a poor indicator of an adsorbent's capacity to remove organic micropollutants from natural water. For all four surface chemistry modifications, ACF10 performed significantly better than ACF15 and ACF20 for the removal of TCE from natural water, a result that concurred with the trends observed in ultrapure water (Table 4.5).

Figure 4.28 shows the effect of ACF pore structure on the removal of MTBE from SJDW for the acid-washed ACFs. For MTBE, AW10 performed significantly worse than AW15 and AW20. Furthermore, the IAST model could not describe the adsorption behavior of AW10 especially at intermediate adsorbent doses. Because the IAST model is only capable of modeling direct site competition, it is likely that NOM adsorption on AW10 caused pore blockage and/or pore constriction that prevented MTBE from accessing a portion of the pore structure over the tested 4-week equilibration time. One possible explanation is that NOM adsorbed at the openings of the pores thus narrowing the pore entrances. Because MTBE approximates the shape of a tetrahedron, it would be more difficult for an MTBE molecule to enter this constricted pore entrance than for a TCE molecule, which has a flat configuration. The ability of the IAST to describe the MTBE adsorption data on AW10 at the largest carbon doses suggests that the competition mechanism shifted from pore constriction/blockage at high NOM loadings to direct site competition at lower NOM loadings.

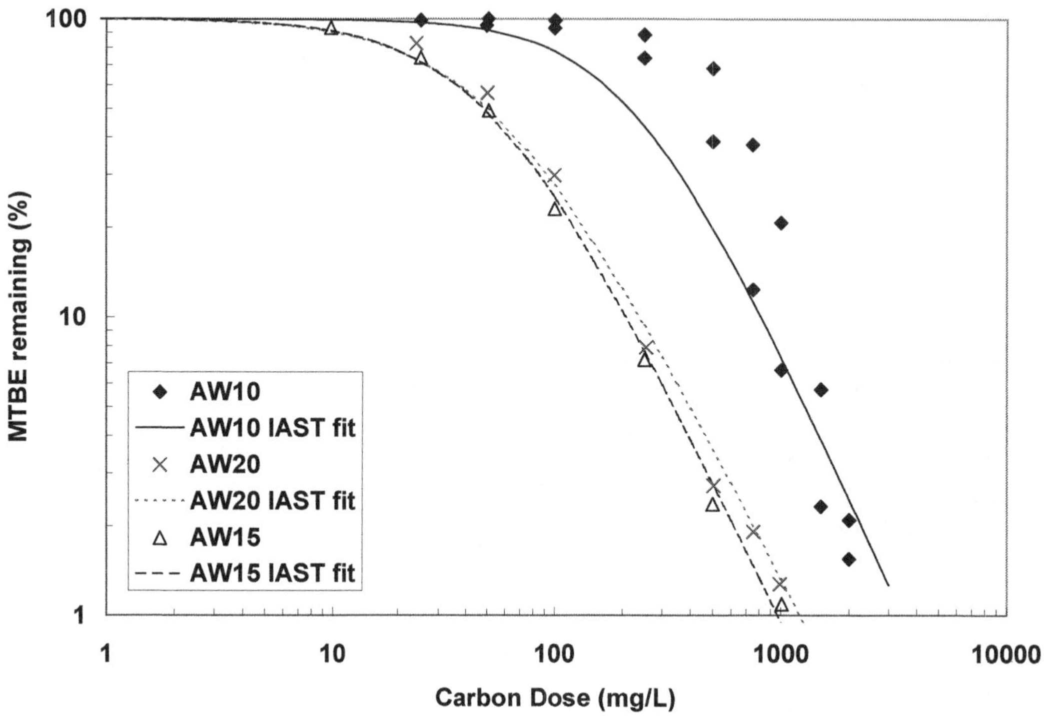

Figure 4.28 Effect of ACF pore structure on MTBE adsorption from SJDW (acid-washed ACFs)

Figure 4.29 shows MTBE isotherms collected in SJDW for HAW10 and OAW10. As was observed for AW10, the IAST model was not capable of predicting the adsorption behavior of HAW10, and a similar observation was made for AAW10 (not shown in Figure 4.29). However, the model was able to describe MTBE adsorption on OAW10. Because the acidic OAW10 carried a negative surface charge at the pH of SJDW, repulsive electrostatic forces likely decreased NOM adsorption at the pore entrances. In contrast, the remaining ACFs exhibited less negative or positive surface charges at the pH of SJDW according to pH_{pzc} data presented in Table 4.2. Consequently, electrostatic forces would have favored NOM adsorption on the latter ACFs. Another possible explanation is that the acidic nature of OAW10 caused the water near the carbon surface to be acidic. Consequently, NOM may have condensed into a coiled form and diffused deeper into the adsorbent pores, where it competed directly for adsorption sites. Even though pore constriction/blockage could not be inferred for OAW10, the oxidized carbon was not as effective for MTBE adsorption from SJDW (Figure 4.29), a result that illustrates that water adsorption in addition to NOM adsorption determines the effectiveness of an adsorbent for the removal of micropollutants from natural water.

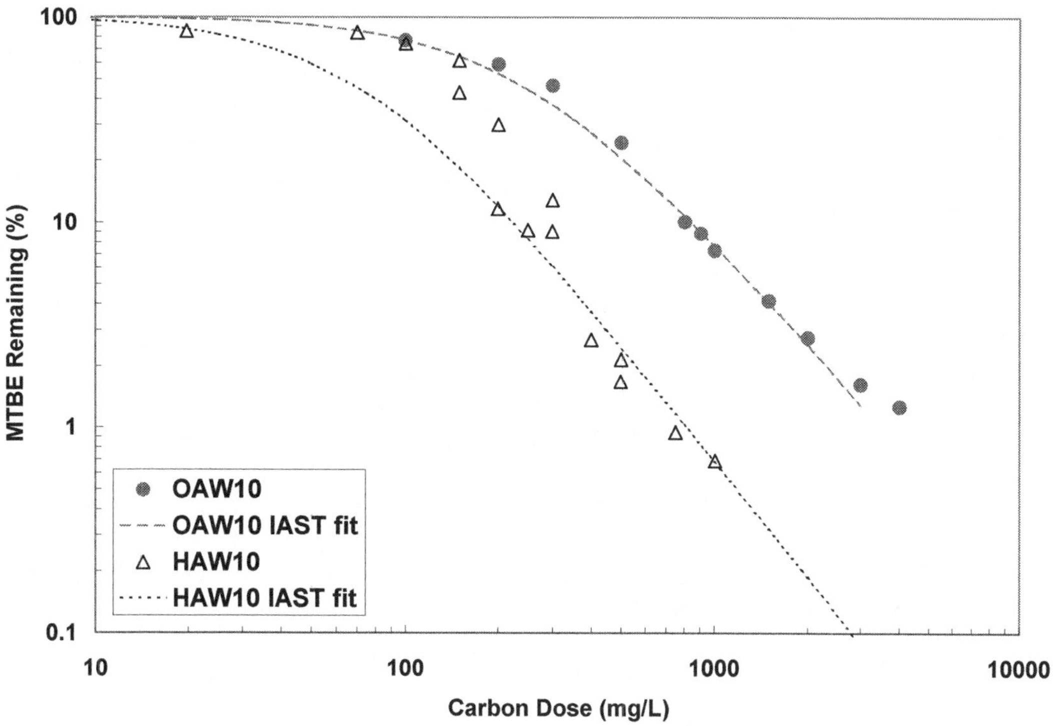

Figure 4.29 Effect of ACF surface chemistry modifications on MTBE adsorption from SJDW (ACF10)

To assess in which pore sizes TCE and MTBE adsorb preferentially in the presence of NOM, adsorption capacities at an equilibrium liquid-phase concentration of 10 µg/L (q_{e10}) were correlated with ACF pore volumes in a given size range for each surface chemistry modification. Following a trial-and-error approach similar to that described by Ebie et al. (1995), the results showed that TCE adsorption was controlled by pores in the 7-9 Å pore width range while MTBE adsorption was controlled by pores in the 8-11 Å pore width range. For TCE, Figure 4.30 illustrates that the available pore volume in the 7-9 Å pore width range primarily controlled TCE adsorption on the more hydrophobic ACFs (AW, HAW, AAW) although TCE adsorption capacities of hydrogen-treated ACFs were consistently larger than those of acid-washed and ammonia-treated ACFs. Single-solute TCE adsorption capacities correlated best with the volume of pores in the 7-10 Å pore width range. The most likely explanation for this change is that NOM out-competed TCE in pores with pore widths in the 9-10 Å range or that TCE adsorbed relatively uninhibited in pores in the 7-9 Å pore width range. Figure 4.30 also shows that the correlation established by the more hydrophobic ACFs predicted the adsorption capacity of the relatively hydrophobic F600 and G219 GACs reasonably well. For the oxidized ACFs, a separate correlation indicated that surface chemistry effects in addition to the availability of a suitable pore volume controlled TCE adsorption for ACFs with larger oxygen contents (~10%). The correlation established by the oxidized ACFs appropriately predicted the adsorption capacity of Picazine GAC with an oxygen content of about 16%.

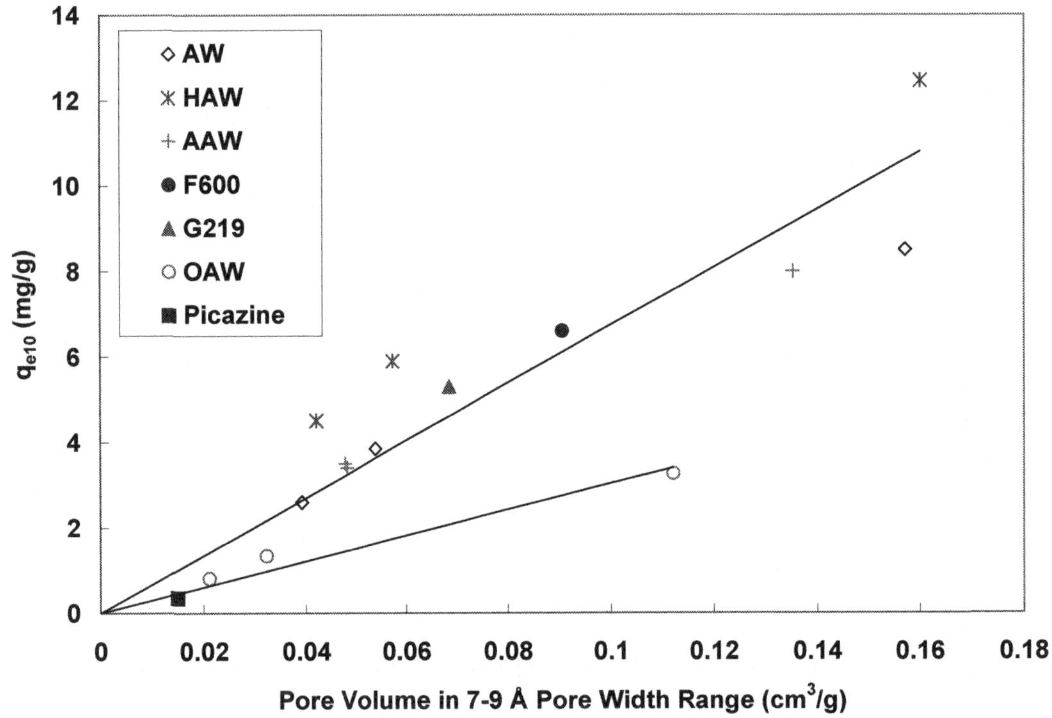

Figure 4.30 Effect of pore volume on TCE adsorption from SJDW

As shown in Figure 4.31, a plot of MTBE adsorption capacity versus micropore volume in the 8-11 Å pore width range showed a trend similar to that for TCE; i.e., one correlation described MTBE adsorption on the more hydrophobic ACFs while a separate correlation described MTBE adsorption on oxidized ACFs. The results depicted in Figure 4.31 suggest that MTBE adsorption was controlled by the pore volume in the 8-11 Å pore width range, which is the same pore size range that controlled MTBE adsorption in single-solute systems (Figures 4.10). This result suggests that MTBE had to compete with NOM over the entire pore size range that controls MTBE adsorption.

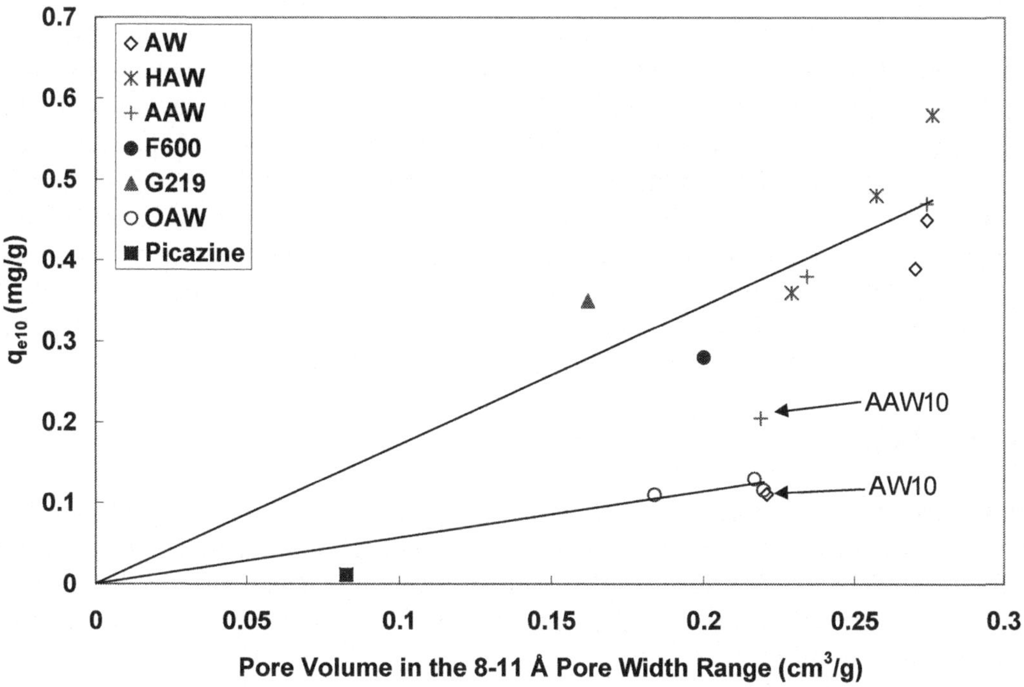

Figure 4.31 Effect of pore volume on MTBE adsorption from SJDW

Two hydrophobic adsorbents that deviated from the trend established by the hydrophobic ACFs were AW10 and AAW10, a result that was likely attributable to pore blockage effects. Given that MTBE adsorption occurred primarily in pores with pore widths in the 8-11 Å range and given that the dominant pore sizes of ACF10 were in the 7-11 Å width range, it appears reasonable that even a small quantity of adsorbed NOM caused pore blockage or constrictions in pore size that precluded MTBE access to adsorption sites. MTBE isotherms in SJDW were also poorly described by the IAST, a result that further suggests that strongly adsorbing NOM components caused pore blockage or constrictions at pore entrances. As a result of pore blockage, MTBE adsorption capacities (q_{e10}) of AW10 and AAW10 in SJDW were 69 and 64% smaller, respectively, than the single-solute MTBE adsorption capacities. These capacity reductions were the most severe among the adsorbents and suggest that pore blockage is especially detrimental to the adsorption of micropollutants from natural water, a result that concurs with the findings of Pelekani and Snoeyink (1999).

Although the results in Figure 4.29 suggest that NOM adsorption also caused some pore blockage on HAW10, reductions in MTBE adsorption capacity were not as severe as those observed for AW10 and AAW10. Possibly, the lower polarity of HAW10 reduced the severity of pore blockage by lowering the quantity of adsorbed NOM at pore entrances or by reducing the combined effects of water and NOM adsorption on pore blockage. Electrostatic effects most likely did not play a role in reducing pore blockage effects on HAW10 given that its surface charge should be intermediate to those of AW10 and AAW10 based on pH_{pzc} data (Table 4.2).

Pore blockage/constriction effects could not be inferred from MTBE adsorption data collected in SJDW with ACF15 and ACF20, adsorbents whose dominant pore sizes were in the 9 to 13 Å width range. Therefore, these results suggest that the adverse effects of pore blockage by NOM can be avoided if the micropore size distribution extends to pore widths that are approximately twice the kinetic diameter of the adsorbate molecule. It needs to be emphasized that seemingly small changes in micropore size distributions (7-11 Å for ACF10, 9-13 Å for ACFs 15 and 20; i.e., a shift of 2 Å) have important consequences on the adsorption of trace organic contaminants from natural water. The micropore size distribution of an activated carbon not only determines whether a large pore volume in a suitable pore size range for the target compound is available but also the mechanism by which NOM competes with trace organic contaminants for adsorption sites [see also related studies by Pelekani and Snoeyink (1999, 2000, 2001)].

Surface Chemistry Effects

Figure 4.32 illustrates the effect of surface chemistry on TCE adsorption for ACF20. TCE adsorption capacities increased with increasing adsorbent hydrophobicity, and similar results were observed for ACFs 10 and 15. These results confirm the trends described for single-solute systems and illustrate that hydrophobic adsorbents retain their advantage over more polar adsorbents in the presence of NOM.

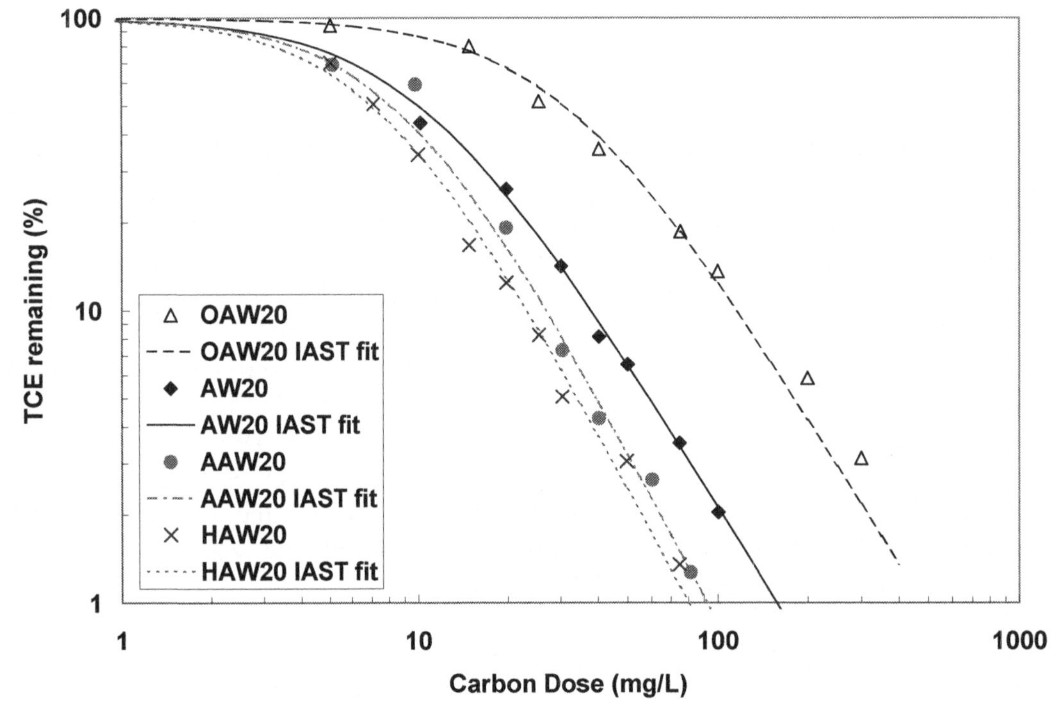

Figure 4.32 Effect of ACF surface chemistry modifications on TCE adsorption from SJDW (ACF20)

The effects of surface chemistry on the adsorption of MTBE by ACF20 are shown in Figure 4.33. As was observed for TCE, the more hydrophobic carbons (AW20, HAW20, and AAW20) adsorbed MTBE more effectively than the more hydrophilic carbon (OAW20). Again, these results are similar to those obtained in the absence of NOM and confirm that enhanced water adsorption on polar adsorbents interferes with the adsorption of both nonpolar and polar adsorbates from aqueous solution. Enhanced water adsorption on polar surface sites therefore negated the potential benefit of providing specific adsorption sites (i.e. oxygen containing functional groups such as carboxylic acid or phenolic groups) capable of forming hydrogen-bonds with MTBE. Surface chemistry modifications of ACF15 yielded trends similar to those observed for ACF20 in Figure 4.33.

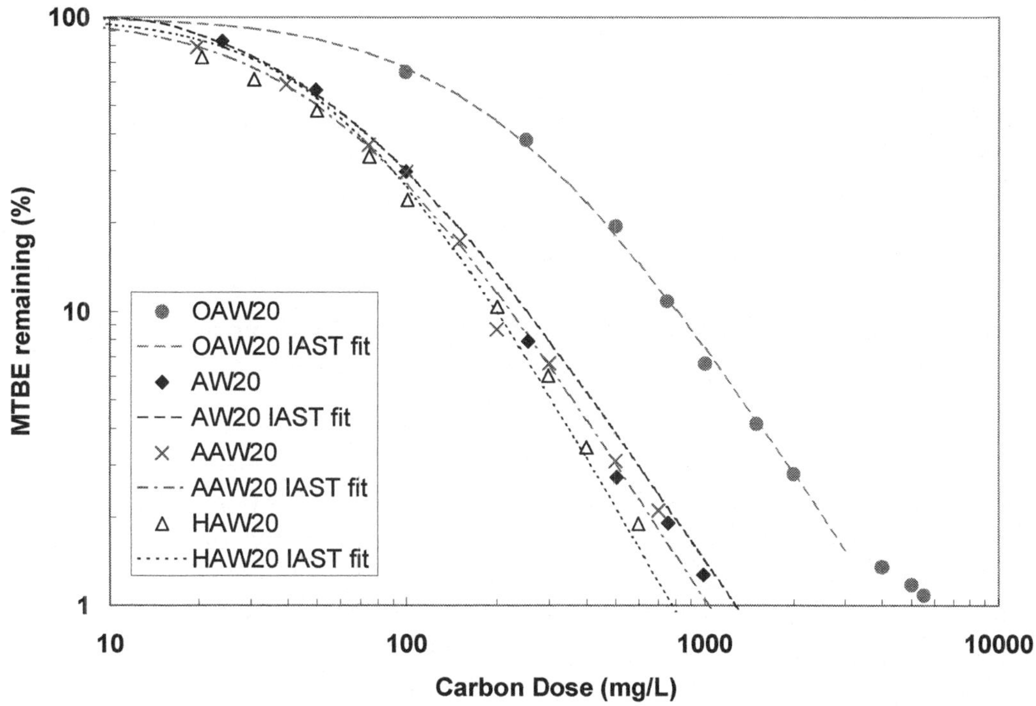

Figure 4.33 Effect of ACF surface chemistry modifications on MTBE adsorption from SJDW (ACF20)

However, surface chemistry modifications of ACF10 yielded a different trend for MTBE adsorption from SJDW. Figure 4.34 illustrates that OAW10 performed similarly or better than AW10. As explained earlier, a likely explanation for this result is that pore blockage/restriction occurred on AW10 but not on OAW10 as a result of repulsive electrostatic forces between the negatively charged surface of OAW10 and negatively charged NOM. However, Figure 4.29 above illustrated that HAW10 exhibited a larger MTBE adsorption capacity in SJDW than OAW10 once the mechanism of competition shifted from pore blockage/constriction to direct competition on HAW10.

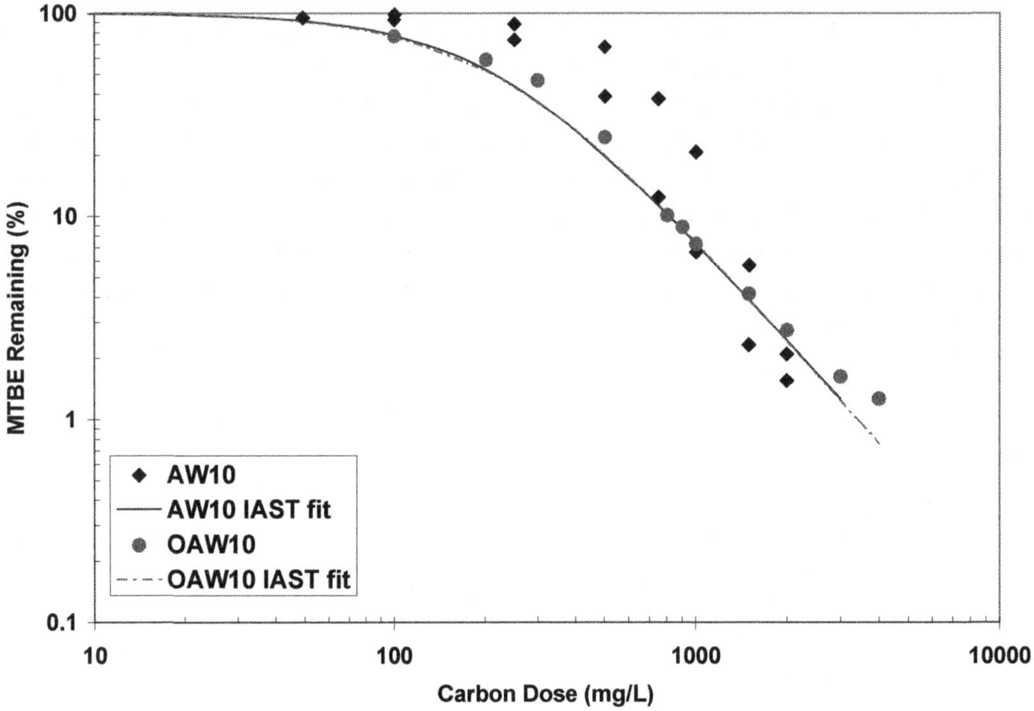

Figure 4.34 Effect of ACF surface chemistry modifications on MTBE adsorption from SJDW (AW10 vs. OAW10)

To determine the effects of adsorbent polarity on TCE and MTBE adsorption in the presence of NOM, adsorption capacities at an equilibrium liquid-phase concentration of 10 μg/L (q_{e10} in mg/g) were first normalized by the pore volume in the pore width range that controlled the adsorption of the targeted micropollutant (i.e., 7-9 Å for TCE, 8-11 Å for MTBE). The thus normalized micropollutant adsorption capacities (q'_{e10} in mg micropollutant/cm^3 pore volume) were then plotted against the sum of the oxygen and nitrogen contents (mmol/g) of each adsorbent as shown in Figures 4.35 and 4.36 for TCE and MTBE, respectively. Figure 4.35 shows that the normalized TCE adsorption capacity decreased with increasing adsorbent polarity, and a similar trend was observed for MTBE in Figure 4.36.

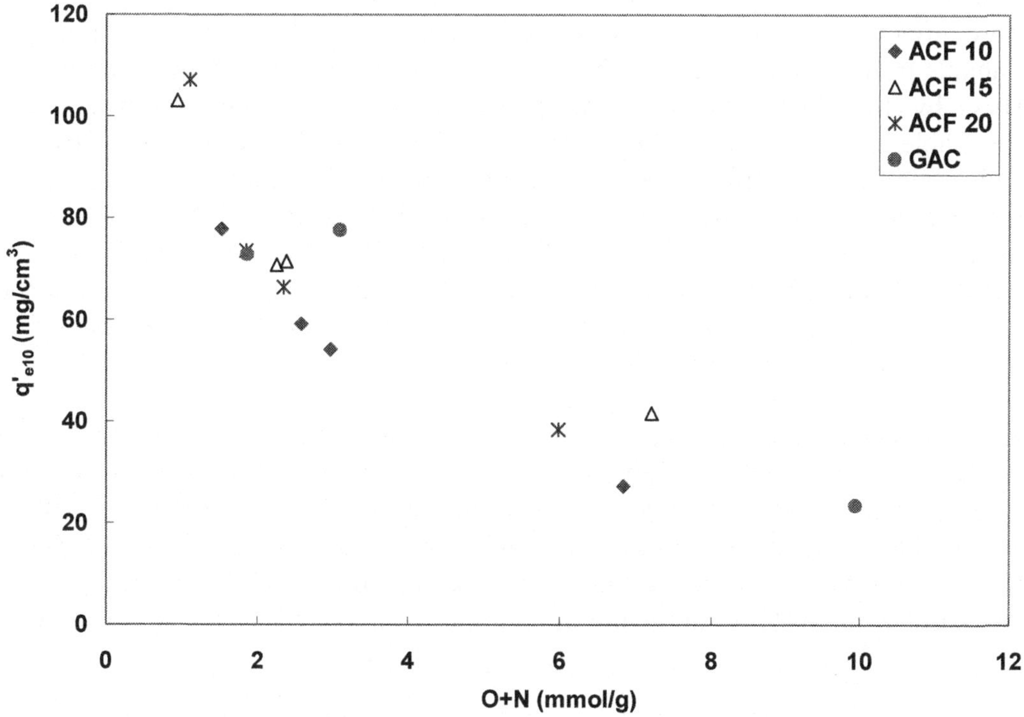

Figure 4.35 Effect of adsorbent hydrophobicity on TCE adsorption capacity in the presence of NOM ($C_{0, TCE}$ = 100 µg/L)

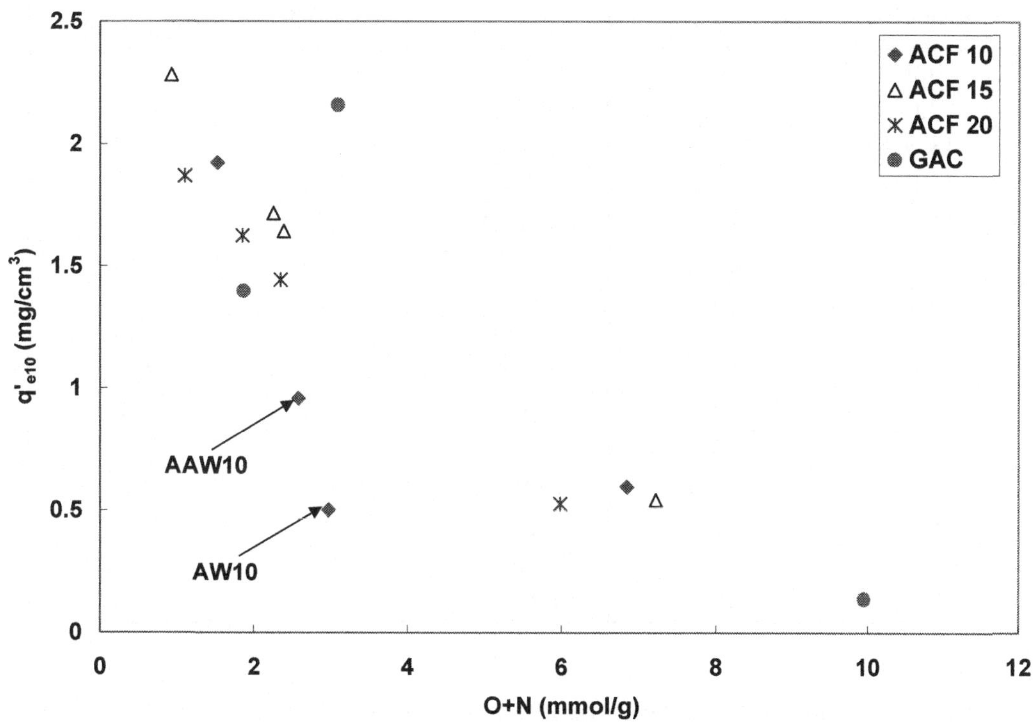

Figure 4.36 Effect of adsorbent hydrophobicity on MTBE adsorption capacity in the presence of NOM ($C_{0, MTBE}$ = 100 µg/L)

These results illustrate that hydrophobic adsorbents; i.e., adsorbents with a low heteroatom content, are more effective for the removal of both relatively hydrophobic and relatively hydrophilic pollutants from natural water. Furthermore, Figures 4.35 and 4.36 illustrate that the trends established by the ACF matrix appropriately described the performance of the commercially available GACs. Overall, the results in Figures 4.35 and 4.36 suggest that micropollutant adsorption is most effective on adsorbents that are hydrophobic [(O+N) < 2 to 3 mmol/g] and exhibit a large volume of pores with a diameter that corresponds to approximately 1.5 times the kinetic diameter of the adsorbate. Furthermore, MTBE adsorption data from SJDW suggests that the detrimental effects of pore blockage as a result of NOM adsorption can be avoided if the micropore size distribution extends to pore widths that correspond to approximately twice the kinetic diameter of the adsorbate.

To illustrate the importance of adsorbent selection, Table 4.7 ranks the adsorbents in terms of the carbon dosage required to remove 90% of either TCE or MTBE from SJDW in the order from best to worst. The dosages are based on IAST model results and are in close agreement with the experimental data in the carbon dosage range necessary to remove 90% of TCE or MTBE. In addition, Table 4.7 summarizes the percentage reduction in adsorption capacity (q_{e10}) that resulted from the presence of NOM when the initial micropollutant concentration was 100 μg/L.

Table 4.7 Adsorbent performance rankings in the presence of NOM

TCE			MTBE		
Carbon	**Reduction in q_{e10} (%)**	**Carbon Dose for 90% removal**	**Carbon**	**Reduction in q_{e10} (%)**	**Carbon Dose for 90% removal**
HAW10	30.9%	8.3	HAW15	22.6%	156
AW10	47.0%	11.6	HAW20	40.4%	199
AAW10	43.3%	12.6	AAW15	32.3%	200
F600	43.8%	15.1	AW15	25.9%	207
HAW15	31.7%	17.5	HAW10	32.2%	222
G219	33.9%	22.0	AAW20	22.8%	222
HAW20	45.1%	23.0	AW20	19.5%	240
AW15	48.9%	26.0	G219	20.4%	255
AAW20	47.5%	26.8	F600	39.2%	315
AAW15	53.4%	27.2	AAW10	64.0%	460
OAW10	31.7%	30.1	OAW15	26.3%	790
AW20	57.5%	38.0	OAW20	16.3%	800
OAW15	45.3%	71.0	AW10	68.7%	809
OAW20	69.6%	115.0	OAW10	37.1%	825
Picazine	14.4%	283.0	Picazine	14.7%	7900

To achieve 90% TCE removal from SJDW, adsorbent doses of about 8 to 300 mg/L were required, and the rankings were affected both by adsorbent pore structure and surface chemistry. The most suitable adsorbents for TCE were hydrophobic adsorbents that exhibited a large pore volume in the 7 to 9 Å range width range, i.e. HAW10, AW10, and AAW10. Among the commercially available GACs, F600 most closely met the requirements for a suitable adsorbent for TCE removal from natural water, and this adsorbent was nearly 20 times as effective as the polar Picazine. Percentage reductions in TCE adsorption capacites were generally in the range of 30 to 50% and showed little dependence on pore structure or surface chemistry given that none of the studied adsorbents exhibited pore size distributions that caused pore blockage/constriction effects that affected TCE access to adsorption sites in suitable micropores. Table 4.8 ranks the studied adsorbents in terms of TCE and MTBE adsorption capacities based on the single-solute isotherm data in the order from best to worst. Because the effects of pore structure and surface chemistry had little effect on the competitive effect of NOM, the adsorbent rankings for TCE removal from single-solute systems were similar to those for TCE removal from SJDW.

Table 4.8 Adsorbent performance rankings in ultrapure water

TCE		MTBE	
Carbon	**Single-solute q_{e10} (mg/g)**	**Carbon**	**Single-solute q_{e10} (mg/g)**
HAW10	18.0	HAW15	0.814
AW10	16.0	HAW20	0.805
AAW10	14.1	AAW15	0.694
F600	11.7	HAW10	0.648
HAW15	8.64	AW15	0.607
HAW20	8.20	AAW10	0.597
G219	8.02	AAW20	0.492
AW15	7.53	AW20	0.484
AAW15	7.29	F600	0.460
AAW20	6.66	G219	0.439
AW20	6.12	AW10	0.355
OAW10	4.77	OAW10	0.175
OAW20	2.67	OAW15	0.160
OAW15	2.45	OAW20	0.139
Picazine	0.409	Picazine	0.0134

To achieve 90% MTBE removal from SJDW, adsorbent doses of about 160 to 8000 mg/L were required. MTBE adsorption was most effective with hydrophobic adsorbents that exhibited a large pore volume in the 8 to 11 Å width range and whose micropore widths extended to about 13 Å to prevent pore blockage. Thus, the most effective adsorbents for MTBE were hydrophobic ACFs 15 and 20 (i.e., hydrogen-treated, acid-washed, and ammonia-treated) whose pores were concentrated in the 9-13 Å width range. Among the commercially available GACs, coconut-shell-based G219 was most effective for MTBE removal from natural water (dominant pore widths: 7-13 Å), while the polar Picazine GAC was again the least effective. In terms of capacity reduction percentage, NOM adsorption had the most detrimental effects on the MTBE adsorption capacities of AAW10 and AW10, a result that was primarily attributable to pore blockage or pore constriction as described above. While capacity reduction percentages were lowest for polar adsorbents with larger micropores (i.e. OAW15, OAW20, or Picazine), a result consistent with the electrostatic repulsion between negatively charged adsorbent surfaces and negatively charged NOM, polar adsorbents still performed poorly in natural water because of the detrimental effects of enhanced water adsorption. Rankings of the adsorbents in terms of MTBE removal effectiveness from SJDW (Table 4.7) were generally similar to those observed in ultrapure water (Table 4.8) with two notable exceptions; the effectiveness of both AAW10 and AW10 decreased greatly because of the detrimental effects of pore blockage or restriction on the MTBE adsorption capacity.

Adsorbents were also ranked in terms of BET surface area and micropore volume (Table 4.9). A comparison of the rankings in Table 4.9 with those presented in Tables 4.7 and 4.8 clearly illustrates that BET surface area and micropore volume are not related to the adsorption capacity of trace organic compounds in single-solute systems and in the presence of competing NOM.

Table 4.9 Adsorbent rankings in terms of BET surface area and micropore volume

BET surface area		Micropore volume	
Carbon	**BET surface area (m^2/g)**	**Carbon**	**Micropore volume (cm^3/g)**
HAW20	1710	HAW20	0.622
AAW20	1710	AAW20	0.599
Picazine	1680	AW20	0.594
AW20	1670	HAW15	0.593
HAW15	1530	AAW15	0.573
OAW20	1530	AW15	0.567
AAW15	1510	OAW20	0.504
AW15	1480	OAW15	0.439
G219	1460	G219	0.438
OAW15	1350	HAW10	0.414
F600	1290	AW10	0.387
HAW10	830	Picazine	0.377
AW10	760	AAW10	0.371
OAW10	750	F600	0.353
AAW10	720	OAW10	0.340

CONCLUSIONS

To evaluate pore structure and surface chemistry effects on the adsorption of TCE and MTBE from aqueous solution, a matrix of activated carbon fibers (ACFs) with three activation levels and four surface chemistry levels was studied along with three commercially available GACs. From the relationships observed between the adsorbent characteristics and the micropollutant adsorption isotherm data, the following conclusions were drawn:

- Regardless of the pore structure or surface chemistry, adsorbents always exhibited a larger adsorptive capacity for TCE than for MTBE. Smaller MTBE adsorption capacities were principally, but not solely, attributable to the greater aqueous solubility of MTBE.
- The pore volume of micropores with widths corresponding to about 1.3 to 1.8 times the kinetic diameter of the targeted adsorbate controlled the single-solute adsorption capacity. Consequently, relatively small changes in the micropore size distribution of an adsorbent (e.g., 7-11 Å widths in ACF 10 versus 9-13 Å widths in ACFs 15 and 20) can alter the effectiveness of an adsorbent for a given micropollutant. Similarly, small differences in the kinetic diameters among individual target adsorbates (e.g., 5.6 Å for TCE vs. 6.2 Å for MTBE) affect the choice of the most effective adsorbent. In natural water, the pore volume of micropores with widths corresponding to about 1.5 times the kinetic diameter of the target pollutant controlled the adsorption capacity. However, if the dominant pore size range and the range of pore sizes, in which the micropollutant adsorbs, have similar upper bounds, pore blockage or restriction can occur as a result of NOM adsorption. Consequently, adsorption capacities decreased greatly as was observed for MTBE adsorption from SJDW on AW10 and AAW10. To prevent pore blockage, the dominant micropore range should be extended to pore widths that are approximately twice the kinetic diameter of the target adsorbate.
- In conjunction with pore size distribution data, adsorbent polarity, as expressed by the sum of the oxygen and nitrogen contents (mmol/g) has the potential to serve as a useful activated carbon selection criterion. As the polarity of a carbon increases, or the O+N content increases, the micropollutant adsorption capacity decreases. This result is consistent with enhanced water adsorption on polar surface functional groups, which leads to the formation of water clusters that block micropollutant access to adsorption sites. The detrimental effects of water adsorption on polar adsorbents was observed for both hydrophobic (TCE) and hydrophilic (MTBE) adsorbates. Based on the results of this study, an effective adsorbent for the removal of organic contaminants from aqueous solution should have and O+N content of less than 2 to 3 mmol/g.
- Attempts to minimize the adverse effects of NOM adsorption by selecting carbons with a negative surface charge (i.e. oxidized ACFs) were unsuccessful because the carbons contained a larger concentration of polar functional groups. The detrimental effects of enhanced water adsorption on these carbons outweighed any benefits that were derived from minimizing NOM adsorption through repulsive electrostatic forces.

Overall, the results of this study showed that the adsorption capacity for a micropollutant is maximized for adsorbents with

- a large volume of micropores with widths that are about 1.5 times larger than the kinetic diameter of the target adsorbate,
- a micropore size distribution that extends to widths that are approximately twice the kinetic diameter of the target adsorbate to prevent pore blockage, and
- a hydrophobic pore surface chemistry which, when expressed as the sum of the oxygen and nitrogen contents, should not exceed 2 to 3 mmol/g.

CHAPTER 5
RECOMMENDATIONS TO UTILITIES

When facing the task of selecting a suitable activated carbon for the removal of trace organic compounds from drinking water sources, water treatment professionals are faced with the problem of having to choose from a wide array of activated carbons that are marketed by numerous manufacturers. Frequently, the selection process is based on the cost of activated carbon (e.g., \$/ton) or on descriptors such as the iodine number or Brunauer, Emmett, and Teller (BET) surface area (e.g., \$/(mg/g) of iodine number or \$/(m^2/g) of BET surface area). The results of this study showed that none of the above selection criteria are valid for optimizing the selection process. E.g., for the activated carbon fiber (ACF) matrix studied in Chapter 4, AW10 with a BET surface area of 760 m^2/g exhibited a TCE adsorption capacity that was about 3 times larger than that of AW20 with a BET surface area of 1670 m^2/g. Also, among the commercial GACs that were evaluated in this study, the GAC with the largest BET surface area exhibited the lowest adsorption capacity for TCE and MTBE because its micropores were too large and its surface too polar. Similarly, the iodine number, which correlates well with BET surface area, is an inappropriate activated carbon selection criterion.

To optimize the selection process for activated carbons, utilities need to consider (1) the adsorption capacity of a given adsorbent for a given adsorbate, (2) the rate at which the contaminant is removed from the water by the adsorbent, and (3) the potential interference from water treatment chemicals, such as coagulants or oxidants. The focus of this study was to identify factors that affect adsorption capacity, which include

- the size of the targeted pollutant
- the pore size distribution of the activated carbon, and
- the hydrophobicity of the activated carbon.

With respect to adsorbate size, the results of this study indicate that the kinetic diameter of the targeted pollutant determines the pore size range that should be maximized in the activated carbon. Information about kinetic diameters of organic pollutants is scattered in the literature, however. As a reasonable substitute, the paper by Okouchi et al. (1992) may be consulted, which lists the total surface area (TSA) of many organic pollutants. A molecular diameter can be estimated from the TSA values presented by Okouchi et al. (1992) by assuming that the molecules are spheres [i.e., $d = (TSA/\pi)^{0.5}$].

With respect to the adsorbent pore size distribution, it would be useful for activated carbon manufacturers to provide such information in the format shown in Figure 4.3 of this report. It should be realized, however, that a standard method for the determination of micropore size distributions is lacking to date. The density functional theory (DFT) allows the calculation of micropore size distributions, and it is now frequently included with the software packages of gas adsorption analyzers used to characterize activated carbons. However, questions remain with respect to which adsorbate (e.g., N_2, Ar, CO_2) is most suitable for the determination of micropore size distributions. Perhaps even more importantly, the calibration procedure used in the development of DFT kernels will affect the pore size distribution that results from the DFT. E.g.,

DFT kernels exist that were calibrated with Vulcan carbon black while others were calibrated with Sterling graphite. A standard method for determining micropore size distributions is therefore required before utilities can compare micropore size distributions from different sources in a reliable manner. For the DFT kernel employed in this study, the adsorbent pore volume should be maximized at a pore width that is approximately 1.5 times larger than the kinetic diameter of the targeted adsorbate. Furthermore, to minimize pore blockage effects by natural organic matter (NOM) adsorption, pore size distributions should extend to widths that correspond to approximately twice the kinetic diameter of the targeted adsorbate.

With respect to the hydrophobicity of activated carbons, the results of this study showed that for carbons with similar pore size distributions, hydrophobic carbons, i.e., carbons with low oxygen and nitrogen contents, are most effective for the removal of trace organics from natural water. This result was valid for both hydrophobic (TCE) and hydrophilic (MTBE) adsorbates and could be explained by the preferential adsorption of water on polar surface functional groups. Measuring the elemental composition of an adsorbent is probably the simplest way to obtain information about the hydrophobicity of an adsorbent. To maximize the adsorption capacity, an adsorbent should exhibit a low ash content as well as a low content of polar heteroatoms; i.e., the sum of an adsorbent's oxygen and nitrogen content, should be less than about 2 to 3 mmol/g. Again, providing information about the elemental composition and ash content of adsorbents would be useful information that should be provided by activated carbon manufacturers. However, the elemental composition and ash content can also be obtained with relative ease by submitting adsorbent samples to commercial laboratories.

While a large adsorption capacity is important for selecting effective activated carbons, rapid adsorption kinetics can be of even greater importance when contact times are short (e.g., Newcombe et al., 2002). In that case, the mesopore volume of an activated carbon may become an additional selection criterion. Recommendations regarding a minimum mesopore volume are not yet possible because adsorption kinetics were not measured in this study.

Overall, the results of this study emphasize that selection criteria such as the unit cost of activated carbon (e.g., \$/ton), iodine number, or BET surface area (e.g., \$/(mg/g) of iodine number or \$/(m^2/g) of BET surface area) are inappropriate for selecting an activated carbon that can meet a given treatment goal at minimum cost. Currently, it is still difficult to write bid specifications that are solely based on activated carbon pore structure and surface chemistry although the above recommendations can serve as initial guidelines. In the meantime, utilities should rely on performance-based tests to determine the anticipated treatment cost with a given activated carbon. Especially when searching for cost-effective powdered activated carbons (PAC), utilities should rely on bench-scale tests to determine the required carbon dose to meet a given treatment goal under conditions that are applicable to their plant (contact time, mixing conditions, sequence of chemical additions). Treatment costs can then be estimated from the results of such tests by multiplying the required PAC dose by the unit cost of the PAC [i.e., Treatment cost (\$/day) = Required PAC dose to meet treatment goal (mg/L) * PAC cost (\$/lb) * daily flow (mgd) * 8.34 (lb/Mgal)/(mg/L)].

REFERENCES

Adams JQ and Clark RM. Cost estimates for GAC treatment systems. Journal of the American Water Works Association 1989; 81(1): 35-42.

Aiken GR, Brown PA, Noyes TI, Pinckney DJ. Molecular size and weight of fulvic and humic acids from the Suwannee River. In Averett RC, Leenheer JA, McKinight DM Thorn KA, editors. Humic Substance in the Suwannee River, Georgia: Interactions, Properties, and Properties. Denver, CO: United States Geological Survey Water-Supply Paper 2373. 1994.

Bansal RC, Donnet JB, Stoeckli F. Active carbon. New York: Marcel Dekker, Inc. 1988.

Barton SS, Evans MJB, Holland J, Koresh JE. Water and cyclohexane vapour adsorption on oxidized porous carbon. Carbon 1984; 22(3):265-272.

Barton SS, Evans MJB, Halliop E, MacDonald JAF. Acidic and basic sites on the surface of porous carbon. Carbon 1997; 35(9):1361-1366.

Belfort G. Selective adsorption of organic homologues onto activated carbon from dilute aqueous solutions. Solvophobic interaction approach and correlations of molar adsorptivity with physicochemical parameters. Environ. Sci. Technol. 1979; 13(8):939-946.

Bernazeau F, Mandra V, Charles P, Anselme C, Bersillon J. Pesticides removal of activated carbon: competitive adsorption with natural organic matter. Water Supply 1996; 14(2):43-48.

Biniak S, Szymanski G, Siedlewski J, Swiatkowski A. The characterization of activated carbons with oxygen and nitrogen surface groups. Carbon 1997; 35(12):1799-1810.

Bjelopavlic M, Newcombe G, Hayes R. Adsorption of NOM on activated carbon: Effect of surface charge, ionic strength, and pore volume distribution. Journal of Colloid and Interface Science 1999; 210:271-280.

Boehm HP, Diehl E, Heck W, Sappok R. Surface oxides of carbon. Angew. Chem. Internat. Edit. 1964; 3:669-677.

Boehm HP, Voll M. Basische Oberflächenoxide auf Kohlenstoff – I. Adsorption von Säuren. Carbon 1970; 8:227-240.

Boehm HP. Some aspects of the surface chemistry of carbon blacks and other carbons. Carbon 1994; 32:759-769.

Bolto B, Abbt-Braun G, Dixon D, Eldridge R, Frimmel F, Hesse S, et al. Experimental evaluation of cationic polyelectroletes for removing natural organic matter from water. Water Science and Technology 1999; 40(9):71-79.

Cambridge Structural Database. CSD Version 5.20. Cambridge UK: Cambridge Crystallographic Data Centre.

Carter MC, Kilduff JE, Weber WJ Jr. Site energy distribution analysis of preloaded adsorbents. Environ. Sci. Technol. 1995; 29(7):1773-1780.

Chen G, Dussert W, Suffet H. Evaluation of granular activated carbons for removal of methylisoborneol to below odor threshold concentration in drinking water. Water Research 1997; 31(5):1155-1163.

Chin, YP, Aiken G., O'Loughlin E. Molecular weight, polydispersity, and spectroscopic properties of aquatic humic substances. Environmental Science and Technology 1994; 28(11):1853-1858.

Chintawar PS, Greene HL. Adsorption and catalytic destruction of trichloroethylene in hydrophobic zeolites. Applied Catalysis B: Environmental 1997; 14:37-47.

Considine R, Denoyel R, Pendleton P, Schumann R, Wong SH. The influence of surface chemistry on activated carbon adsorption of 2-methylisoborneol from aqueous solution. Colloid and Surfaces A: Physicochemical and Engineering Aspects 2001; 179:271-280.

Coughlin RW, Ezra FS. Role of surface acidity in the adsorption of organic pollutants on the surface of carbon. Environ. Sci. Technol. 1968; 2(4):291-297.

Coulson CA. The electronic structure of the boundary atoms of a graphite layer. Proc. of the fourth conf. on carbon, Oxford: Pergamon Press, 1960: 215-219.

Crittenden JC, Sanongraj S, Bulloch JL, Hand DW, Rogers TN, Speth, TF, Ulmer M. Correlation of aqueous-phase adsorption isotherms. Environ. Sci. Technol. 1999; 33(17):2926-2933.

Daley MA, Tandon D, Economy J, Hippo EJ. Elucidating the porous structure of activated carbon fibers using direct and indirect methods. Carbon 1996; 34(10):1191-1200.

Davis SW, Powers SE. Alternative sorbents for removing MTBE from gasoline-contaminated ground water. J. Environ. Engrg. 2000; 126(4):354-360.

Derbyshire F, Jagtoyen M, Andrews R, Rao A, Martin-Gullon I, Grulke EA. Carbon materials in environmental applications. In Radovic LR, editor. Chemistry and Physics of Carbon, Vol. 27, New York: Marcel Dekker, 2001: 1-66.

Derylo-Marczewska A, Jaroniec M, Gelbin D, Seidel A. Heterogeneity effects in single-solute adsorption from dilute solutions on solids. Chemica Scripta 1984; 24:239-246.

Dimotakis ED, Cal MP, Economy J, Rood MJ, Larson SM. Chemically treated activated carbon cloths for removal of volatile organic carbons from gas streams: evidence for enhanced physical adsorption. Environ. Sci. Tech. 1995; 29(7):1876-1880.

Dubinin MM, Zaverina ED, Serpinsky VV. The sorption of water vapor by active carbon. J. Am. Chem. Soc. 1955; 77:1760-1766.

Dubinin MM. The potential theory of adsorption of gases and vapors for adsorbents with energetically nonuniform surfaces. Chem. Rev. 1960; 60: 235-241.

Ebie K, Li F, Hagishita, T. Effect of pore size distribution of activated carbon on the adsorption of humic substances and trace organic compounds. Water Supply 1995; 13(3/4):65-70.

Ebie K, Li F, Azuma Y, Yuasa A, Hagishita T. Pore distribution effect of activated carbon in adsorbing organic micropollutants from natural water. Water Research 2001; 35(1):167-179.

Franz M, Arafat HA, Pinto NG. Effect of chemical surface heterogeneity on the adsorption mechanism of dissolved aromatics on activated carbon. Carbon 2000; 38(13):1807-1819.

Garten, VA, Weiss DE. A new interpretation of the acidic and basic structures in carbons. II. The chromene-carbonium couple in carbon. Austr. J. Chem. 1957; 10:309-328.

Gillogy TET, Snoeyink VL, Elarde JR, Wilson CM, Royal EP. ^{14}C-MIB adsorption on PAC in natural water. Journal of the American Water Works Association 1998; 90(1):98-108.

Graham MR, Summers RS, Simpson MR, Macleod BW. Modeling equilibrium adsorption of 2-methylisoborneol and geosmin in natural waters. Water Research 2000; 34(8):2291-2300.

Greene BC, Snoeyink VL, Pogge FW. Adsorption of Pesticides by powdered activated carbon. American Water Works Association: Denver, CO. 1994.

Haist-Gulde B. Zur Adsorption von Spurenverunreinigungen aus Oberflächenwässern. Doctoral Dissertation, University of Karlsruhe, Germany. 1991.

Haist-Gulde B, Baldauf G, Brauch, HJ. Models and predictability of the micropollutant removal by adsorption on activated carbon. In Hutzinger O, editor. The Handbook of Environmental Chemistry, Vol. 5, Part B. Berlin, Germany: Springer Verlag. 1995:129-138.

Hopman R, Siegers WG, and Kruithof JC. Organic micropollutant removal by activated carbon fiber filtration. Water Supply 1995; 13 (3/4):257-261.

Janssen RJJ, van Bekkum H. Amination and ammoxidation of activated carbons. Carbon 1994; 32(8):1507-1516.

Janssen RJJ, van Bekkum H. XPS of nitrogen-containing functional groups on activated carbon. Carbon 1995; 33(8):1021-1027.

Kaneko Y, Abe M, Ogino K. Adsorption characteristics of organic compounds dissolved in water on surface-improved activated carbon fibers. Colloids and Surfaces 1989; 37:211-222.

Karanfil T, Kitis M, Kilduff JE, Wigton A. Role of granular activated carbon surface chemistry on the adsorption of organic compounds. 2. Natural Organic Matter. Environmental Science and Technology. 1999; 33(18):3225-3233.

Karanfil T, Kilduff JE. Role of granular activated carbon surface chemistry on the adsorption of organic compounds. 1. Priority pollutants. Environ. Sci. Technol. 1999; 33(18): 3217-3224.

Kasaoka S, Sakata Y, Tanaka E, Naitoh R. Preparation of activated fibrous carbon from phenolic fabric and its molecular-sieve properties. International Chemical Engineering 1989a; 29(1):101-114.

Kasaoka S, Sakata Y, Tanaka E, Naitoh R. Design of molecular-sieve carbon. Studies on the adsorption of various dyes in the liquid phase. International Chemical Engineering 1989b; 29(4):734-742.

Kilduff JE. Polyelectrolyte adsorption by activated carbon and the effects of preloading on trichloroethylene adsorption. Ph.D. Dissertation. The University of Michigan: Ann Arbor, MI. 1995.

Kilduff JE, Karanfil T, Chin Y, Weber WJ. Adsorption of natural organic polyelectrolytes by activated carbon: A size exclusion chromatography study. Environmental Science and Technology 1996; 30: 1336-1343.

Kilduff JE, Karanfil T, Weber WJ Jr. TCE adsorption by GAC preloaded with humic substances. J. AWWA 1998; 90(5):76-89.

Kilduff JE and Wigton A. Sorption of TCE by humic-preloaded activated carbon: Incorporating size-exclusion and pore blockage phenomena in a competitive adsorption model. Environmental Science and Technology 1999; 33(2):250-256.

Knappe DRU, Matsui Y, Snoeyink VL. Predicting the Capacity of Powdered Activated Carbon for Trace Organic Compounds in Natural Waters. Environmental Science and Technology 1998; 32(11):1694-1698.

Lambert JB. Organic Structural Spectroscopy. Upper Saddle River, NJ: Prentice Hall. 1998.

Lee MC, Snoeyink VL, Crittenden JC. Activated Carbon Adsorption of humic substances. Journal of the American Water Works Association 1981; 73(8):440-446.

Leon y Leon CA, Radovic LR. Interfacial chemistry and electrochemistry of carbon surfaces. In Thrower PA, editor. Chemistry and Physics of Carbon, Vol. 24, New York: Marcel Dekker, 1992:213-310.

Leon y Leon CA, Solar JM, Calemma V, Radovic LR. Evidence for the protonation of basal plane sites on carbon. Carbon 1992; 30(5): 797-811.

Lundelius EF. Adsorption und Löslichkeit. Koll. Zeitschr. 1920; 26(4):145-151.

Mahajan OP, Moreno-Castilla C, Walker PL Jr. Surface-treated activated carbon for removal of phenol from water. Sep. Sci. and Technol. 1980; 15(10):1733-1752.

Mangun CL, Benak KR, Daley MA, Economy J. Oxidation of activated carbon fibers: effect on pore size, surface chemistry, and adsorption properties. Chem. Mater. 1999; 11:3476-3483.

Mangun CL, Benak KR, Economy J, Foster KL. Surface chemistry, pore sizes and adsorption properties of activated carbon fibers and precursors treated with ammonia. Carbon 2001; 39:1809-1820.

Mattson JS and Mark HB Jr. Activated carbon - surface chemistry and adsorption from solution. New York: Marcel Dekker. 1971.

McCallum CL, Bandosz TJ, McGrother SC, Müller EA, Gubbins KE. A molecular model for adsorption of water on activated carbon: comparison of simulation and experiment. Langmuir 1999; 15:533-544.

McGuire MJ, Suffet IH, Radziul JV. Assessment of unit processes for the removal of trace organic compounds from drinking water. J. AWWA 1978; 70(10):565-572.

Meldrum BJ, Rochester CH. In situ infrared study of the surface oxidation of activated carbon dispersed in potassium bromide. J. Chem. Soc. Faraday Trans. 1990; 89(11):2997-3002.

Menendez JA, Phillips J, Xia B, Radovic LR. On the modification and characterization of chemical surface properties of activated carbon: in the search of carbons with stable basic properties. Langmuir 1996a; 12(18):4404-4410.

Menendez JA, Radovic LR, Xia B, Phillips J. Low-temperature generation of basic carbon surfaces by hydrogen spillover." J. Phys. Chem. 1996b; 100(43):17243-17248.

Morris G and Newcombe G. Granular activated carbon: The variation of surface properties with the adsorption of humic substances. Journal of Colloid and Interface Science 1993; 159:413-420.

Müller G, Radke CJ, Prausnitz JM. Adsorption of weak organic electrolytes from dilute aqueous solution into activated carbon. 1. Single-solute systems. Journal of Colloid and Interface Science 1985; 103(2):466-483.

Müller U, Hess F, Worch E. Impact of organic matter adsorbability on micropollutant removal by activated carbon. Journal of Water Supply Research and Technology-Aqua 1996; 45(6):273-280.

Müller EA, Rull LF, Vega LF, Gubbins KE. Adsorption of water on activated carbons: a molecular simulation study. J. Phys. Chem. 1996; 100:1189-1196.

Müller EA, Gubbins KE. Molecular simulation study of hydrophilic and hydrophobic behavior of activated carbon surfaces. Carbon 1998; 36(10):1433-1438.

Müller EA, Hung, FR, Gubbins KE. Adsorption of water vapor-methane mixtures on activated carbons. Langmuir 2000, 16:5418-5424.

Myers AL, Prausnitz JM. Thermodynamics of mixed-gas adsorption. AIChE Journal 1965; 11(1):121-127.

Najm, IN, Snoeyink VL, Richard Y. Effect of initial concentration of a SOC in natural water on its adsorption by activated carbon. Journal of the American Water Works Association 1991a; 83(8):57-63.

Najm IN, Snoeyink VL, Lykins BW, Adams JQ. Using powdered activated carbon: A critical review. Journal of the American Water Works Association 1991b; 83(1): 65-76.

Nakahara M, Sanada Y. FT-IR ATR Spectroscopy of the edge surface of pyrolytic graphite and its surface/ PVC interface. J. Mater. Sci. 1995; 30:4363-4368.

Narbaitz RM and Benedek A. Adsorption of 1,1,2-trichloroethane from river water. Journal of Environmental Engineering 1994; 120(6):1400-1415.

Newcombe G and Drikas M. Adsorption of NOM onto activated carbon: Electrostatic and non-electrostatic effects. Carbon 1997; 35(9):1239-1250.

Newcombe G, Drikas M, Hayes R. Influence of characterised natural organic matter on activated carbon adsorption: II. Effect on pore volume distribution and adsorption of 2-methylisoborneol. Water Research 1997; 31(5):1065-1073.

Newcombe G. Charge vs. porosity – Some influences on the adsorption of natural organic matter (NOM) by activated carbon. Water Science and Technology 1999; 40(9):191-198.

Newcombe G, Morrison J, Hepplewhite C, Knappe DRU. Simultaneous adsorption of MIB and NOM onto activated carbon: II. Competitive effects. Carbon 2002, in press.

Nichols EM, Einarson MD, Beadle SC. Strategies for characterizing subsurface releases of gasoline containing MTBE, Publication No. 4699. Washington DC: American Petroleum Institute. 2000.

Noh JS, Schwarz JA. Estimation of the point of zero charge of simple oxides by mass titration. J. Coll. Interf. Sci. 1989; 130(1):157-164.

Okouchi S, Saegusa H, Nojima O. Prediction of environmental parameters by adsorbability index: water solubilities of hydrophobic organic pollutants. Environment International 1992; 18: 249-261.

Pan D, Jaroniec M. Adsorption and thermogravimetruc studies of unmodified and oxidized active carbons. Langmuir 1996; 12:3657-3665.

Pauling L. The nature of the chemical bond, and the structure of molecules and crystals: an introduction to modern structural chemistry; 3rd edition. Ithaca NY: Cornell University Press. 1960.

Pelekani C, Snoeyink VL. Competitive adsorption in natural water: role of activated carbon pore size. Water Research 1999; 33(5):1209-1219.

Pelekani C, Newcombe G, Snoeyink VL, Hepplewhite C, Assemi S, Beckett R. Characterization of natural organic matter using high performance size exclusion chromatography. Environmental Science & Technology 1999; 33(16):2807-2813.

Pelekani C, Snoeyink VL. Competitive adsorption between atrazine and methylene blue on activated carbon: the importance of pore size distribution. Carbon 2000; 38:1423-1436.

Pelekani C, Snoeyink VL. A kinetic and equilibrium study of competitive adsorption between atrazine and Congo red dye on activated carbon: the importance of pore size distribution. Carbon 2001; 39:25-37.

Pendleton P, Wong SH, Schumann R, Levay G, Denoyel R, Rouquerol J. Properties of activated carbon controlling 2-methylisoborneol adsorption. Carbon 1997; 35(8):1141-1149.

Perdue EM and Lytle CR. Distribution model for binding of protons and metal ions by humic substances. Environmental Science & Technology 1983; 17(11):654-660.

Polovina M, Babic B, Kaluderovic B, Dekanski A. Surface characterization of oxidized activated carbon cloth. Carbon 1997; 35(8):1047-1052.

Puri BR. Surface complexes on carbons. In Walker PL Jr., editor. Chemistry and Physics of Carbon, Vol. 6, New York: Marcel Dekker, 1970:191-282.

Puri BR. Physicochemical aspects of carbon affecting adsorption from the aqueous phase. In McGuire MJ and Suffet IH, editors. Advances in Chemistry Series, Vol. 202, Washington DC: American Chemical Society, 1983:77-93.

Qi S, Adham SS, Snoeyink VL, Lykins BW. Prediction and verification of atrazine adsorption by PAC. Journal of Environmental Engineering 1994; 120(1):202-218.

Radke CJ, Prausnitz JM. Thermodynamics of multi-solute adsorption from dilute liquid solutions. AIChE Journal 1972; 18(4):761-768.

Sano T, Hasegawa M, Kawakami K, Yanagishita H. Separation of methanol/methyl tert-butyl ether mixture by pervaporation using silicalite membrane. J. Membrane Science 1995; 107: 193-196.

Schnitzer M, Khan SU. Humic Substances in the Environment. New York: Marcel Dekker, Inc. 1972.

Schwarzenbach RP, Gschwend PM, Imboden DM. Environmental Organic Chemistry. New York: John Wiley and Sons, Inc. 1993.

Sellitti C, Koenig JK, Ishida H. Surface Characterization of Graphitized Carbon Fibers by Attenuated Total Reflection Fourier Transform Infrared Spectroscopy. Carbon 1990; 28:221-228.

Sing KSW, Everett DH, Haul RAW, Moscou L, Pierotti RA, Rouquerol J, Siemieniewska T. Pure and Applied Chemistry 1985; 57:603-619.

Sing KSW. Physisorption of nitrogen by porous materials. Journal of Porous Materials 1995; 2:5-8.

Smisek M, Cerny S. Active carbon. Manufacture, properties and applications. New York: Elsevier Publishing Company, 1970.

Snoeyink VL, Weber WJ Jr. Surface functional groups on carbon and silica. Progress in Surface and Membrane Science, Vol. 5, New York: Academic Press, 1972:63-119.

Snoeyink VL, Summers RS. Adsorption of organic compounds. In Letterman RD, editor. Water Quality and Treatment; 5th edition. New York: McGraw-Hill, Inc. 1999.

Sontheimer H, Crittenden JC, Summers RS. Activated carbon for water treatment. DVGW-Forschungsstelle, Engler-Bunte-Institute, University of Karlsruhe: FRG. 1988.

Starsinic M, Taylor RL, Walker PL Jr., Painter PC. FTIR studies of saran chars. Carbon, 1983: 21(1):69-74.

Summers RS and Roberts PV. Activated carbon adsorption of humic substances I. Heterodisperse mixtures and desorption. Journal of Colloid and Interface Science 1988a; 122(2):367-381.

Summers RS and Roberts PV. Activated carbon adsorption of humic substances II. Size exclusion and electrostatic interactions. Journal of Colloid and Interface Science 1988b; 122(2):382-397.

Weber WJ Jr. Physicochemical processes for water quality control. New York: John Wiley and Sons, 1972, pp 199-259.

IWA Publishing's authorised EU representative for General Product Safety Regulations is Diane D'Arras, 15 rue Duret, 75116 Paris, France, e-mail: safety@iwap.co.uk.

Printed and bound by CPI Group (UK) Ltd, Croydon, CR0 4YY
17/09/2026
02218810-0010